HOW MATHEMATICS HAPPENED

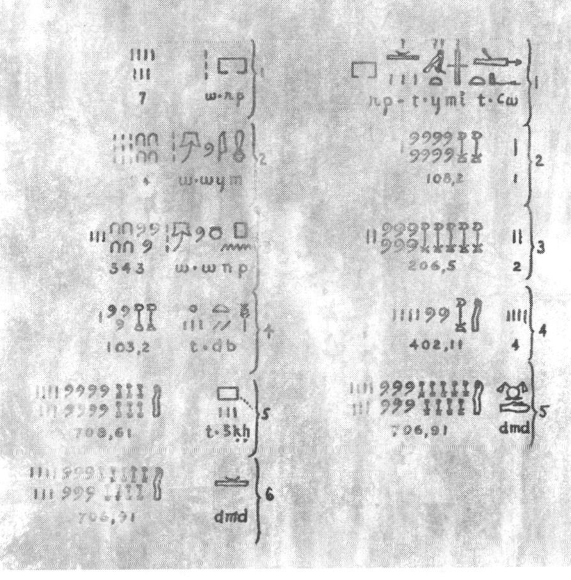

HOW MATHEMATICS HAPPENED
THE FIRST 50,000 YEARS

PETER S. RUDMAN

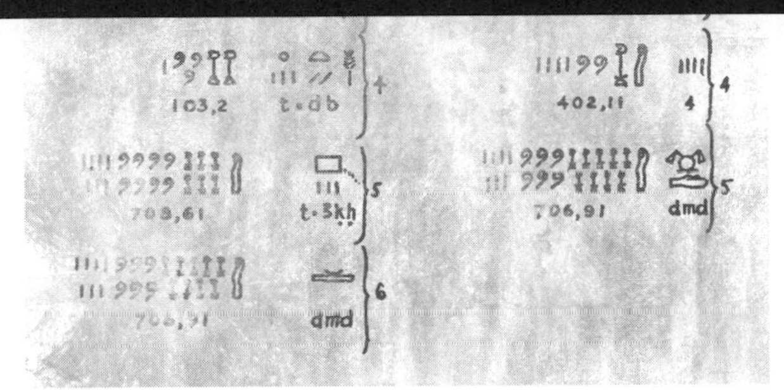

Prometheus Books
59 John Glenn Drive
Amherst, New York 14228-2197

Published 2007 by Prometheus Books

Inquiries should be addressed to
Prometheus Books
59 John Glenn Drive
Amherst, New York 14228–2197
VOICE: 716–691–0133, ext. 207
FAX: 716–564–2711
WWW.PROMETHEUSBOOKS.COM

11 10 09 08 07   5 4 3 2 1

Library of Congress Cataloging-in-Publication Data

Rudman, Peter Strom.
How mathematics happened : the first 50,000 years / by Peter S. Rudman.
p. cm.
Includes bibliographical references and index.
ISBN: 978–1–59102–477–4
1. Mathematics, Ancient. 2. Mathematics, Babylon. 3. Mathematics—Egypt—History. 4. Metrology—Egypt—History. I. Title.

QA22.R86 2006
510.93—dc22

2006020255

Printed in the United States on acid-free paper

To
Pythagoras and the Beatles,
whose "All Is Number" and "All You Need Is Love"
so aptly define the poles about which my world revolves

CONTENTS

LIST OF FIGURES

LIST OF TABLES

PREFACE

I call the eras from about 50,000 BCE to 2000 BCE when humans learned to do practically all the arithmetic we learn nowadays as children, although with different symbols and methods, the *childhood of mathematics*.

The record of *what* mathematics the ancients did is well chronicled and I have not attempted a comprehensive retelling, although I have brought the record up-to-date with some recent important findings. Chronicling *what* happened is the first step in writing history, but much more interesting and important is *why* events occurred. My interest here is *why* ancient cultures did mathematics in a particular way. *Why* did the Egyptians use an apparently bizarre method of expressing fractions? *Why* did the Babylonians use a number system based on multiples of 60? It is to such intriguing, hitherto unanswered questions that I direct my attention.

What is an archeology and translation problem; *why* is more of a mathematics problem. *Why* involves appreciating how someone was thinking thousands of years ago. It might be thought that answering *why* cannot be more than rank speculation, but the Rosetta Stone that enables deciphering how the ancients thought mathematically is simply the realization that we still use the same natural and intuitive mathematical thinking, which is universal and timeless. That is how millions of years of hominid evolution have wired our brains.

Some *whys* are trivial; the decimal system was obviously invented because we have ten fingers. But consider the number 0.333. Is this an exact value, or is it the common fraction 1/3 expressed as a decimal fraction and rounded off? If it is a rounded-off, *nonterminating* decimal fraction 0.333 . . . , was it correctly rounded off? Nonterminating fractions are a nuisance, always were and always will be. Some four thousand years ago, Egyptians and Babylonians each invented different ways of expressing fractions; both systems had the property that they never, well hardly ever, had to encounter nonterminating fractions. Clearly, *why* these ancient cultures did fractions as they did was at least partly to avoid the nuisance of nonterminating fractions. This example nicely illustrates how appreciation of universal and timeless mathematical thinking enables extracting the *whys* of ancient mathematics from the archeological record, providing a deeper understanding of an interesting chapter in the history of intellectual evolution and useful insights into current mathematical practice. As the *whys* get answered, a better understanding emerges of Mayan, Egyptian, and Babylonian mathematics in particular and of the evolution of mathematics in general.

I deserve no credit or blame for any of the translations of ancient documents. With gratitude and awe for their accomplishments, I use the translations that linguists, philologists, epigraphers, archeologists, and historians (whatever they wish to call themselves) have made over the past century or so, but I simplify the translations into modern arithmetic and algebraic notation. However, my analysis of the mathematical record includes two new inputs. First, I pay particular attention to the role of units of measurement. That simple counting was certainly the first step in the evolution of mathematics has led to overlooking that measurements (length, area, volume, weight, and so on) became important in later eras in the childhood of mathematics. Second, I do not treat Egyptian and Babylonian mathematics as independent developments. Rather, I note that similarities between Egyptian and Babylonian mathematics imply considerable intellectual interaction, and I use documents in each country to help interpret documents in the other. Similarly, I use surviving Hindu documents to help interpret some Babylonian mathematics.

The mathematics developed by about 2000 BCE in Egypt and Babylon was the inheritance of the Greeks. Pythagoras and his successor Greek mathematicians added the concept of rigorous proof; mathematics thereby attained maturity, and its childhood and my story ends. However, to illustrate the watershed difference between Babylonian and Greek mathematics, I give a few examples of Greek solutions to some problems Babylonians had previously treated. In the final chapter, I discuss the way ancient Greeks taught mathematics, which suggests how to improve mathematics education today.

By definition, the *childhood of mathematics* cannot contain any truly difficult concepts, and knowledge of arithmetic and some elementary algebra are the only prerequisites to understanding this book. However, to enable understanding of ancient mathematics in a nonsuperficial way, I introduce and explain some simple number-theory concepts where necessary and no prior familiarity is required. Interleaved throughout the book are relevant poetry, biography, history, and humor to minimize unrelieved-till-the-eyes-glaze-over mathematics.

Included are a few rigorous algebraic proofs to provide insight into some ancient procedures and to avoid ever having to use that annoying expression, "it can be shown." If following a rigorous algebraic proof exceeds your attention span, just accept the conclusion and skip the details. In most cases, I uniquely make algebraic proofs and derivations follow a prior arithmetic solution with just the replacement of numbers by letters to emphasize how the algebra is simply a generalization of the arithmetic.

Chapter 5 on Babylonian mathematics is necessarily more algebraic than the rest of the book, because that was Babylon's most interesting and hitherto misunderstood mathematics. However, this *geometric algebra* is easy to visualize and understand in terms of simple diagrams. Even if you have never learned or have long forgotten what a quadratic equation is and how to solve it, you will find insightful new meaning here to "completing the square."

For the most part, *How Mathematics Happened* is an easy read, but as with any book about science or mathematics, some parts require more concentration. The generally given recommendation, which I endorse, is to read such sections casually first and then, if understanding in depth interests you, read them again with more concentration. Chapter 5 is the most difficult, and not just because it is more algebraic. Translations of newly found texts and new translations of previously known texts require reevaluation of much of Babylonian mathematics. As the adage says, "The devil is in the details," and hence some interpretations rely on rather detailed mathematical dissection of translated cuneiform texts.

If you enjoy recreational math and wish to deepen and test your mastery of new concepts, distributed throughout the book are many FUN QUESTIONS. Basic arithmetic can solve most, but a few that require algebra or trigonometry are so noted. FUN QUESTIONS that are critical to understanding the text have ANSWERS directly following them in the text; ANSWERS to all others are in the appendix, and if you do not enjoy solving these problems, they can be skipped without loss of story continuity or essential concepts. The questions range from very easy to very challenging.

I do not refer to references in the text unless the reference itself is part of the narrative. Rather, appended at the end of the book are notes and references for each chapter. Any reader so inclined can trace sources of translations and concepts to their origins. I have found the Internet to be a valuable source for up-to-date information and I include many references to Internet sources, but with realistic evaluation of their value and limitations. The notes also explain how a particular reference relates to my conclusions.

So now, heed the word of the Lord, go forth and multiply (along with other mathematical operations) as done in those good old biblical days.

1
INTRODUCTION

1.1 Mathematical Darwinism

As far as most people know or care, our number system and the arithmetic we do with it descended from Mount Sinai inscribed on the reverse side of the Ten Commandments: "Thou shalt count with the decimal system using the ten symbols 0, 1, 2, 3, 4, 5, 6, 7, 8, 9; and do arithmetic thus and so."

Actually, the currently used *decimal system*, also called the *base-10 system*, evolved in India about two thousand years ago. It was introduced into Europe about one thousand years later in Latin translations of Arabic translations of Hindu texts. Because the Latin translations credited Arabic sources, these symbols became known as Arabic numerals. The symbols have changed so much over the years that neither the original Hindu, nor their Arabic transcriptions, nor the early European transcriptions are easily recognizable in the present symbols. Only after the invention of printing with movable type by Johannes Gutenberg in 1442 did numbers eventually standardize into the symbols almost universally used today. In our politically correct world, it is better to call the symbols Hindu-Arabic, and I shall refer to them so.

In different times and places, many different number systems and arithmetics evolved. The currently used decimal system and arithmetic are just the fittest that have survived by natural selection, but that does not mean that current practice will necessarily be the fittest in the future. The decimal system, which fortuitously exists because we have ten fingers, is not always the best in our electronic calculator/computer age, and other number systems are now widely used. But the decimal system is so deeply embedded in our culture that it will probably never be replaced for everyday use. However, the ubiquitous electronic calculator is rendering obsolete much of the arithmetic currently taught. When was the last time you actually did a pencil/paper multiplication of a multi-digit number by a multi-digit number?

By about 2000 BCE, at least in Egypt and Babylon, humans had learned to do all the basic arithmetic operations, although not with the same symbols or methods used today; had learned some algebra to generalize arithmetic; and had learned some geometry/trigonometry. We learn this mathematics nowadays as children, and so I call the eras up to about 2000 BCE the *childhood of mathematics*.

When development of the individual mirrors development of the species, this is eruditely expressed in biology jargon as "ontogeny recapitulates phylogeny." The mathematical education of a child nowadays does indeed mirror ancient historical progress in mathematics. However, rather than being analogous to any biological process, the mathematics of a child or a culture naturally starts with the easiest and most frequently used operation—counting. The next step is more generalized addition. The other arithmetic operations are just variations of addition, and hence come later.

To relate the other arithmetic operations to addition, and to introduce algebraic notation and jargon, let a and b be *givens* and x be the *unknown* to be calculated.

- **Addition**: $x = a + b$. In current base-10, pencil/paper arithmetic using Hindu-Arabic numerals, we use a memorized addition table.
- **Subtraction:** $x = b - a$, but in practice we usually do $a + x = b$, which asks what must we add to a to obtain b, and we use the same addition table.
- **Multiplication** is just an efficient way of doing successive additions of the same number: $x = ab = ba = b + b \ldots + b$ (a terms) $= a + a \ldots + a$ (b terms). In current pencil/paper arithmetic, we multiply using a memorized multiplication table and tend to think of it as a unique operation, forgetting that the multiplication table had been calculated by successive additions. Some four thousand years ago, Babylonians also did multiplication using multiplication tables, but the Egyptians used a completely different method that obviated the need for multiplication tables. Somewhat ironically, an electronic calculator multiplies the oldest way, by successive additions.
- **Division:** $x = b/a$, but in practice we usually do $xa = b$, which asks what must we multiply a by to obtain b, and we use the same multiplication table. Since multiplication is successive additions of the same number, it is sometimes convenient to think analogously of division as successive subtractions of the same number: $b - xa = 0$, which asks how many successive subtractions of a from b produce zero.

FUN QUESTION 1.1.1: Prove that $ab = ba$.

ANSWER: One way to prove this is diagrammatically:

Nowadays we tend to think abstractly about multiplication, not relating it to a geometrical diagram, but ancient cultures visualized multiplication in terms of the area of a rectangle. The visualization of generating an area as addition of rows followed naturally from the universal experience of generating a plowed field by addition of furrows. In chapter 3, we shall see how this visualization influenced the evolution of measurement units and even a number system. It is interesting and relevant to note that early Greek writing also followed the path of the plow: the first line was written right to left, the second line from left to right, the third line from right to left, and so forth.

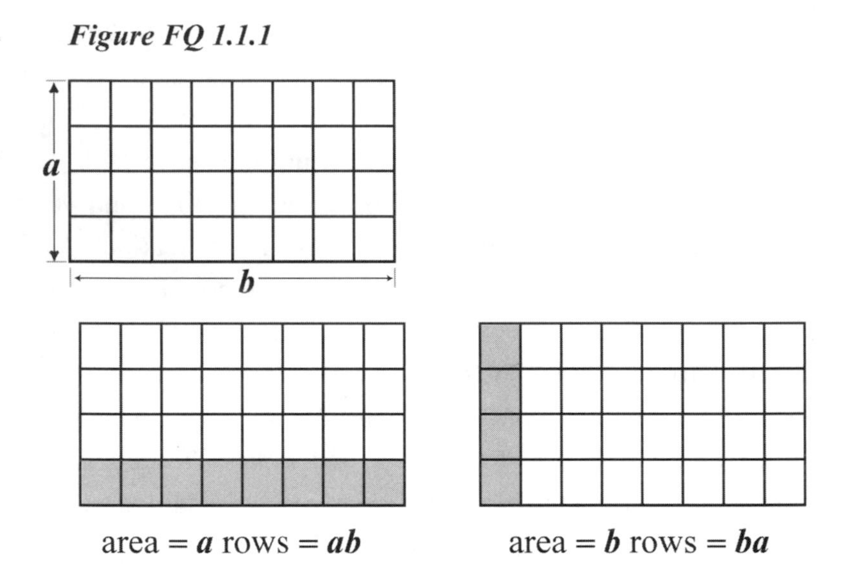

Figure FQ 1.1.1

FUN QUESTION 1.1.2: What is the length of one side of a square that has the same area as a rectangle of sides 2 and 4? This simple but not trivial visualization explains important Egyptian (see section 4.3) and Babylonian (see section 5.3) mathematical inventions.

ANSWER: See the appendix for answers not given contiguously. If you do not enjoy solving these problems, you can skip FUN QUES-

TIONS without contiguous answers with no loss of story continuity or essential concepts.

---

* * * * *

Despite its simplicity and obvious advantages, algebraic notation using a single letter to represent a generalized number is only of rather recent origin, in the sixteenth century. It may appear surprising that neither Euclid (325–265 BCE), nor Archimedes (287–212 BCE), nor any of the other great Greek mathematicians of antiquity invented algebraic notation. A proposed explanation is that the ancient Greeks used alphabetic symbols to write numbers, thereby usurping their use as generalized numbers.

Let us use this example of *why* some Greek mathematician did not invent modern algebraic notation to consider the reliability of *whys* of ancient mathematics. The given explanation appears to be consistent with what is known, and that is always the minimum requirement. Other than that, it is pure speculation. Its reliability depends on the proposer's mathematical intuition. I do not know who proposed this explanation or how good his mathematical intuition was, but my guess is that this *why* is not very reliable because there are other simple alternatives to the usurpation of the Greek alphabet by numbers. For example, Greek mathematicians were certainly aware of other alphabets, such as Latin and Hebrew, that did not use the Greek symbols. Today when mathematicians run out of Latin letters, they use Greek or Hebrew letters, or other symbols such as upside-down or backward letters. Therefore, a guess that appears to me just as probable is that Greek mathematicians were simply preoccupied with other mathematical problems, and their clumsy algebraic notation did not sufficiently bother them to motivate invention of a new notation.

Women's intuition is more hindsight than foresight, as

expressed by a husband's whimsical wish, "If I only knew yesterday, what my wife knows today." But *mathematical intuition* is the starting point of much if not all mathematics. Consider the sums $1 + 3 = 4$, $1 + 5 = 6$, $3 + 5 = 8$, $3 + 7 = 10$. . . . Mathematical intuition leads us to guess that there is some rule operating here. One guess is that there is a rule: the sum of two odd numbers is an even number. In math jargon, such a guess is a *conjecture*. A proof turns a conjecture into a rule. It is easy to prove this conjecture and turn it into a rule.

FUN QUESTION 1.1.3: Prove that the sum of two odd numbers is an even number.

ANSWER: Let m and n be any integers so that $2m$ is an even number. (An even number has a divisor of 2, which $2m$ has, whatever the value of m.) An odd number is then $2m + 1$. The sum of two odd numbers can thus be written as $(2m + 1) + (2n + 1) = 2[m + n + 1]$, an even number, QED. QED is the acronym for *quod erat demonstrandum*, Latin for "which was to be demonstrated." It denotes the end of a successful proof.

However, if you have exceptional mathematical intuition, you may have also noted that the examples, $1 + 3 = 4$, $1 + 5 = 6$, $3 + 5 = 8$, $3 + 7 = 10$ . . . , can be described by another conjecture: every even number is the sum of two prime numbers. (A prime number has only the number 1 and the number itself as divisors.) This intuitive guess is *Goldbach's* (1690–1764) *conjecture*. Since every prime number other than two is an odd number, it might be thought that this would also be easy to prove. No exceptions to this conjecture have ever been found, but neither has anybody been able to prove it. (See sections 5.2 and 6.2 for more on prime numbers.)

The purpose of this discussion of conjectures is to show that there is an essential role for judicious guessing in mathematics, and sometimes that is the best we can do. No document will ever be unearthed in which Archimedes explains why he did not invent algebraic notation. The archeological record only shows *what* was done, never *why* it was done. Answering the *whys* is the primary reason for historical study. I, or anyone else who tries to interpret the archeological record in terms of not just *what* was done, but also in terms of *why* it was so done, must rely on conjecture.

My conjectures are products of my mathematical intuition. I have required them to be consistent with archeological evidence, but also to be consistent with natural and intuitive mathematical thinking that I show is universal and timeless. I believe that I have given not just plausible *whys*, but probable *whys*. Others, with new archeological data, and/or better versed in archeological interpretation, and/or with better mathematical intuition, may be able to produce *whys* that are more probable. But whether my *whys* are right or wrong, it is in the search for them that the fun and mathematical insights reside.

* * * * *

The similarity of the learning sequence of a child to the sequence of historical progress in mathematics is only superficially analogous to any biological process, but the evolution of number systems and arithmetic is truly mathematical Darwinism. An invention that improves a number or arithmetic system is analogous to an enhancing genetic variation in a biological system. In both cases chances are better that the system will survive and propagate.

The game of survival-of-the-fittest mathematics that has been going on for millennia has taken a new twist recently. A new mathematical technique called *genetic programming* generates new solutions by mathematical Darwinism. Rather than limiting solutions to methods that our brains are capable of inventing, random variations in existing solutions generate new solutions. The fittest are selected for survival by defining criteria for what is a better solution. The analog of sexual reproduction is the generation of an offspring solution that combines properties of parent solutions. The analog of mutation is random changes in parts of solutions. Of course, a computer generates this mathematical evolution.

The success of genetic programming is an interesting validation of Darwin's (1809–1882) theory of evolution, and the evolution of the brain in particular. Mathematician Alan Turing (1913–1954), one of the seminal thinkers in computer science, invented the genetic programming concept in 1950, although it was only implemented in the 1990s. In considering the question of whether a computer could ever duplicate a human's brain, Turing conjectured that computer capability would eventually have to increase by the same mechanism that the capability of the human brain has increased, by genetic variation and survival of the fittest.

1.2 THE REPLACEMENT CONCEPT

In order to understand the nondecimal number systems that play an important role in ancient and modern arithmetic, I briefly introduce some simple mathematical concepts in this and the following section.

The finger counting we all first used as children assigns a value of one to each finger, and the count is the sum of extended fingers. This simple *additive* number system, limited to counting up to ten, is possibly the natural and intuitive way humans first started to count.

FUN QUESTION 1.2.1: In *additive* finger counting where all fingers have a value of one, represent an extended finger by the symbol 1 and a nonextended finger by the symbol 0. How many ways can the number one be exhibited? How many ways can the number nine be exhibited? How many ways can the number two be exhibited?

ANSWER: There are ten ways of exhibiting the number one:

10000 00000 01000 00000 00100 00000 00010 00000 00001 00000

00000 10000 00000 01000 00000 00100 00000 00010 00000 00001

See the appendix for remaining answers.

This is a very redundant counting system, which means that many recognizably different finger arrangements, or equivalently symbolic arrangements, have not been exploited to count higher. Clearly, it must be possible to count higher than ten with just ten fingers.

One method of counting higher without adding more fingers is to let each finger on the right hand still have a value of one, but to let each finger on the left hand have a value of five. Each time you count five on the right hand, *replace* the five extended fingers on the right hand by just one extended finger on the left hand (a 1-for-5 replacement). Now you can count to thirty (five on the right hand plus twenty-five on the left hand) with just ten fingers. Whether a finger is on the right or left hand determines its value, and so such finger counting is *positional*. A positional system is also called a *place-value* system. It is implicit in the definition of positional systems that the values at different positions are added. In this example, whatever its position on a given hand, a finger's value is the same, and the value of the hand is calculated additively. To be more precise, this counting system is both additive and positional.

FUN QUESTION 1.2.2: Consider the additive-positional finger-counting system where all fingers on the right hand have a value of one but all fingers on the left hand have a value of five. Again, represent an extended finger by the symbol 1, and a nonextended finger by the symbol 0. How many ways can the number six be exhibited? How many ways can the number twenty-four be exhibited?

ANSWER: There are twenty-five ways of exhibiting the number six. For each exhibition of the left hand, there are five ways of exhibiting the right hand: 10000, 01000, 00100, 00010, 00001. Since there are the same five possible exhibitions of the left hand, the total number of exhibitions is $5 \times 5 = 25$. See the appendix for the other answer.

This solution is an example of the *multiplication rule* of *combinatorial mathematics*, which states that if one arrangement can be done in M ways and another arrangement can be done independently in N ways, then both together can be done in MN ways. This is a very useful rule that I shall use again to solve other problems.

FUN QUESTION 1.2.3: Give a value of one to each finger on the right hand and a value of five only to the first finger on the left hand; assign values to the four remaining fingers on the left hand to maximize the counting limit. What is the maximized counting limit?

ANSWER: The maximized counting limit is 160. If you do not get this answer, see the appendix.

FUN QUESTION 1.2.4: Define a positional finger-counting system such that the value of the first finger on the right hand is one and the value of the second finger is two. Thus, the representation

00000 00001 has a value of one; 00000 00010 has a value of two; 00000 00011 has a value of three; and so forth. Note how the positional property minimizes the number of symbols required because the same symbol has a different value in different positions. Continue the assignment of values to each finger to define a nonredundant finger-counting system. What is the highest that can be counted on ten fingers with this system? What replacement numbers describe this system? Can any finger-counting system that only relies on finger arrangements count higher?

ANSWER: This system can count to 1,023 on just ten fingers! If you have obtained this answer, you have just reinvented the *binary*, also called the *base*-2, number system. See the appendix for solution details.

The ancient solution to the limitations of finger counting was pebble counting. In its simplest realization, pebble counting requires little memorization and uses only the principle of one-for-one correspondence. Consider a shepherd putting a pebble in a bowl as each sheep goes out to pasture and removing a pebble as each sheep returns. The number of pebbles left in the bowl is then the number of sheep lost. This is arithmetic without a need for names for numbers or an ability to articulate counting, with no limit to the number counted, and with a permanent record—a significant advance beyond finger counting.

With growing understanding of the concept of number, our generic shepherd (now perhaps many generations later) wants to know how many sheep he has. A bowl of hundreds of pebbles is not very defining, so he replaces each group of ten pebbles with one larger pebble. Why ten? Because he had first learned to count using his fingers and therefore ten is a natural choice. If he still has too many pebbles to define his number of sheep conveniently, he can replace every ten big pebbles with a bigger pebble, and so on.

There are now never more than nine pebbles of the same size. Unaware though he surely is, he has now invented an *additive*, base-10 number system. The smallest pebble has a value of one; the next larger pebble has a value of ten, the next larger pebble after that has a value of a hundred, and so on. A sequence of replacements using the same replacement number defines a number system with a *base* equal to the replacement number. Figure 1.2.1 illustrates this unrealized and unintended pebble-counting invention of an *additive*, *base-10 number system.*

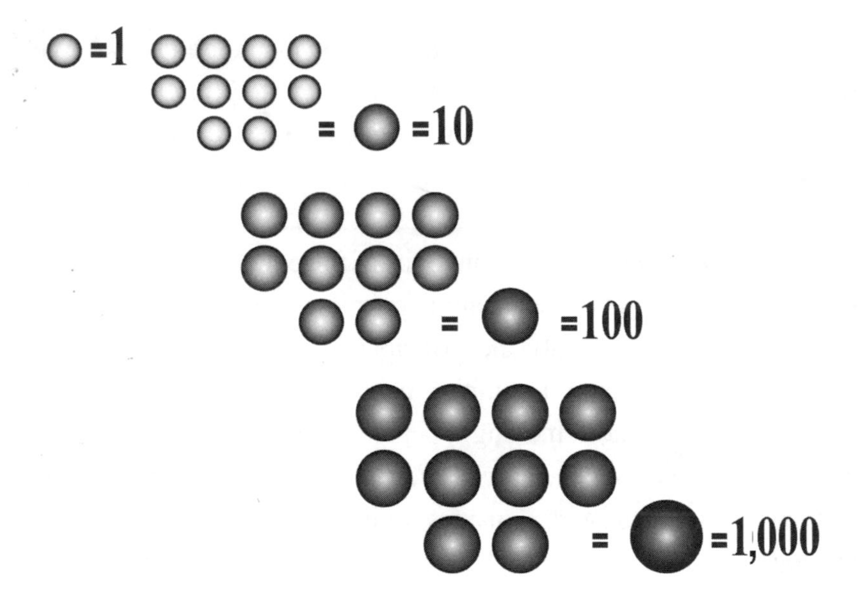

Figure 1.2.1 *Pebble counting with 1-for-10 replacements*

If rather than using pebbles of different sizes as 1-for-10 replacements, ten pebbles in one bowl were replaced by one pebble in a different bowl, a *positional*, *base-10 number system* would have been invented.

Even if one starts with an additive system with different-sized pebbles, for easier visualization of quantity one naturally and intuitively tends to gather each size in a separate pile. Thereby, the values of pebbles are redundantly defined both by pebble size and by pebble position. Eventually this unnecessary redundancy tends to be realized, and hence additive systems tend to evolve into positional systems.

If hands were humankind's first calculating machine, then pebbles were the second. Number systems with a base originated in a pebble-counting era. Once a replacement is chosen for some logical reason, such as 1-for-10 because we have ten fingers, then why change it in a sequence of replacement numbers? Since a sequence of the same replacements defines a number system with a base, number systems with a base tend to evolve naturally once use of replacement has begun. The advantage of a number system with a base becomes more apparent when arithmetic other than simple addition of small numbers is required.

We have now developed the three important characteristics of an efficient number system, and they are valid whether the counting is by fingers, pebbles, or written symbols:

1. **Replacement**—extends the counting limit.
2. **Base**—enables efficient arithmetic.
3. **Position**—minimizes the number of different symbols.

As just noted, there is a logical and real tendency for the invention of replacement to evolve into number systems that have a base and are positional, but it has been a slow evolutionary process. In Egypt, the transition from an additive to a positional number system never occurred over the thousands of years of existence of the Egyptian empire. In Babylon, the additive-to-positional transition did not occur until long after writing had been invented. Fortunately for us, the evolution from additive pebble counting to a written, positional number system with a base that occurred in Babylon over the period from about 9000 to 2000 BCE is archeologically documented in remarkable detail.

While additive number systems were still being used in both Egypt and Babylon, the positional concept was already clearly understood and exploited in abacus addition. However, I shall postpone consideration of the interesting question of why the Egyptians never progressed to a positional number system until section 3.2, after the Egyptian and Babylonian number systems will have been defined in more detail.

The ancient abacus was not the type we are familiar with in recent history, where a fixed number of beads slide along rods. This type of abacus was invented in China only in the eleventh century. Rather, the ancient abacus was simply a surface divided into columns defining positions and with numbers in each position displayed additively with markers (pebbles). Pencil/paper arithmetic gradually replaced such abacus arithmetic in Europe only after the introduction of Hindu-Arabic numerals in the twelfth century. Table 1.2.1 illustrates a decimal-system addition of two addends using an ancient abacus. The operation is simple: whenever there is a sum of ten or more markers in any column, *replace* ten markers by one marker in the next column. Such abacus use thus requires only the application of the same natural and intuitive concept of replacement that led to the unintended invention of a number system with a base.

Such abacus addition requires no memorization whatsoever, just an ability to count to ten. The replacement operation on an abacus is equivalent to the *carry* operation in modern pencil/paper addition. The name *abacus* possibly is derived from a Semitic word meaning dust. For example, in Hebrew the word for dust is *abk*. Thus, an early version of the abacus was possibly dust sprinkled on a surface, and the markers were erased when replaced.

*   *   *   *   *

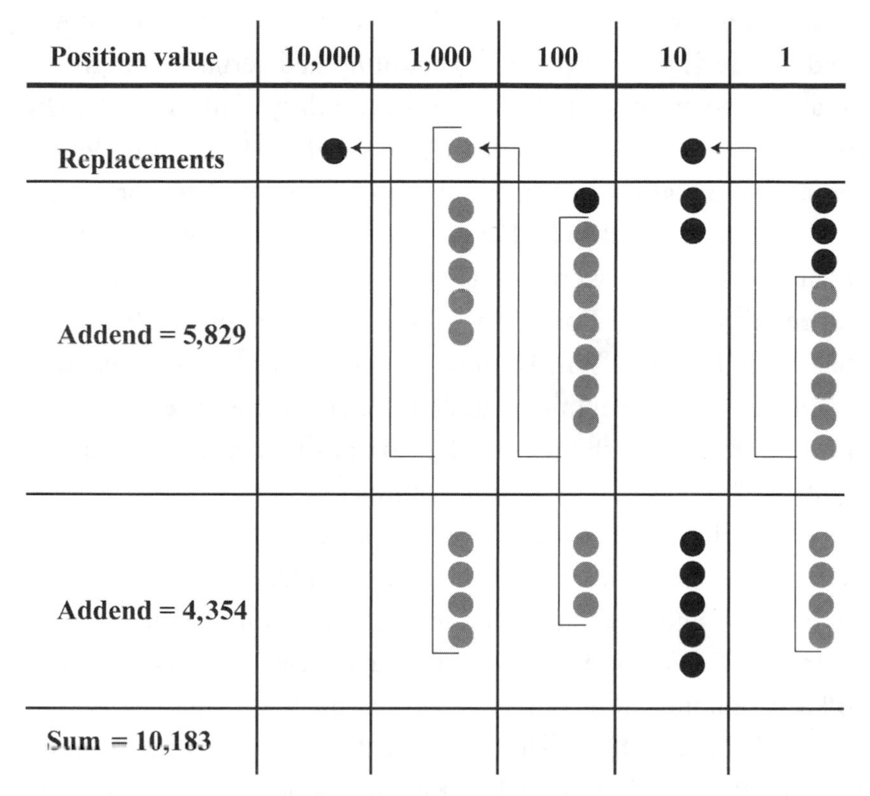

Table 1.2.1 *Abacus, decimal-system addition*

As might be expected, the most popular replacement choice was 1-for-10. A 1-for-20 choice was also popular, showing that when some cultures progressed to requiring counting to higher than ten, at first they simply added toes. Eventually, of course, this was insufficient and counting was extended by the replacement concept. Not illogically, such cultures chose 1-for-20 replacements, and base-20 counting tended to evolve.

We know that the Celts in prehistoric Europe used 1-for-20 replacements in counting because vestiges of its use remain in modern European languages. For example, in French, the word for eighty is *quatre-vingts*, literally four-twenties; the word *quinze-*

vingts, literally fifteen-twenties, survives as a word for three hundred. Celtic vigesimal (base-20) counting had certainly progressed to a two-position system with a vocabulary that could readily express any number up to $(19 \times 20) + 19 = 399$, but no vestige remains of a vigesimal vocabulary for counting higher. For a vigesimal system to count to higher than 399, at least three positions are required with a new word defined to express 20×20, analogous to the definition of the English word *hundred* to express 10×10 in the decimal system. When a number system has not been extended to sufficiently large numbers to define a sequence of replacements, the number system should be referred to as prebase. The Celts certainly had at least a previgesimal number system.

In English, the word *score* has a seldom-used meaning of twenty, as in Lincoln's historic Gettysburg address: "Four score and seven years ago, our fathers brought forth on this continent a new nation. . . ." *Score* originally meant a cut on a stick or a bone in order to keep a numerical record. A longer cut marked twenty, essentially a 1-for-20 replacement, and *score* eventually acquired the meaning of twenty. Up until 1971, when England adopted decimalized currency, the currency denomination of the pound was a 1-for-20 replacement of the shilling. Thus, vestiges of ancient vigesimal or possibly only previgesimal counting also exist in English.

Roman decimal counting did not replace European vigesimal counting because it was palpably better, but because to the victor belongs the spoils and the choice of number system. But for Julius Caesar, we might be counting vigesimally today. Some modern European number systems, such as Welsh (Celtic) and Basque (not Celtic), are called vigesimal because they use vigesimal nomenclature for numbers less than one hundred, but for numbers equal to or greater than one hundred they are unquestionably in decimal. For example, in Welsh, the word for one hundred is *cant*, clearly derived from the Latin *centum*, rather than in vigesimal language as five-twenties; numbers between one hundred and one thousand are defined as multiples of the decimal word *cant*, and the word for one thousand is simply the Latin *mille.* The fact that large-number nomenclature is in decimal hints that ancient European number systems were probably only previgesimal, and very large numbers were probably not named prior to Roman influence.

It might appear strange to refer to the number system of illiterate ancients Celts as positional, but vocalization just as clearly defines positions as does writing. Thus the Celtic vocalization of 315, of which *quinze-vingts quinze* would be the vestige in French, defines a two-position vigesimal number just as the English vocalization, three hundred fifteen, defines a three-position decimal number.

Vigesimal numbers are easily written using the ten Hindu-Arabic numerals instead of the twenty symbols that would be required for a truly vigesimal system. Modern Welsh and Basque use Hindu-Arabic numerals and are indistinguishable from decimal numbers. Thus, Welsh, Basque, and other modern vigesimal users can enjoy using their national languages for number nomenclature and still use the same symbols and arithmetic as the rest of the world.

While ancient Europeans probably never progressed beyond a previgesimal system, in essentially the same centuries, the Maya in Central America did invent a true, twenty-numeral, positional, base-20 number system (see section 3.3).

* * * * *

The mathematician Gottfried von Leibniz (1646–1716) rigorously proved that any integer, except the integer one, could be a base. It is important to have rigorous mathematical proofs such as those by Leibniz so that we unambiguously know the limitations of any concept, but since we are motivated simply by a need for more efficient counting, all number systems for everyday use have evolved naturally and intuitively, starting with the invention of the replacement concept.

1.3 Number Systems

We have now learned that the best number system in any application is probably a positional system with a base. But what base? Let us consider various positional number systems with different bases that are or have been important. We start with our familiar base-10 system. Table 1.3.1 defines aspects of the positional, base-10 number 2,731.

Position	3	2	1	0
Position value	1,000	100	10	1
Position value, power notation	10^3	10^2	10^1	10^0
Digits	2	7	3	1

Table 1.3.1 *Definition of the positional base-10 number 2,731*

The position is the *exponent* in the *power notation* definition of a position's value. Power notation is a simple and useful concept, and I shall frequently employ it. The notation 10^n means the product of n 10s, such as $10^3 = 10 \times 10 \times 10$. The exponent is n. More generally, b^n is the product of n b's. By definition and logically, $b^0 = 1$.

Now let us consider the base-2 system (see Fun Question 1.2.4). As always, the number of symbols equals the value of the base, and so only two symbols are required: 0 and 1. Table 1.3.2 defines aspects of the positional base-2 number 1101.

Position	3	2	1	0
Position value	8	4	2	1
Position value, power notation	2^3	2^2	2^1	2^0
Digits	1	1	0	1

Table 1.3.2 *Definition of the positional base-2 number 1101*

Generalizing the definitions in tables 1.3.1 and 1.3.2, any *N*-position, base-*b*, positional number can be expressed in algebraic notation as a sequence of integers: $a_{N-1}a_{N-2} \ldots a_i \ldots a_2 a_1 a_0$, where the coefficients a_i are integers from 0 to $b - 1$, and the number's value is given as

$$a_{N-1}b^{N-1} + a_{N-2}b^{N-2} \ldots + a_i b^i + \ldots + a_2 b^2 + a_1 b + a_0 \quad (1.3.1)$$

If we employ Hindu-Arabic numerals in equation (1.3.1), we can obtain the decimal value of a number in any base using familiar base-10 arithmetic. The following tabulation presents this calculation for three-digit numbers, $a_2a_1a_0$, for various bases:

Binary (base-2): $101_2 = (1 \times 2^2) + (0 \times 2) + 1 = 4 + 0 + 1 = 5_{10}$

Octal (base-8): $642_8 = (6 \times 8^2) + (4 \times 8) + 2 = 384 + 32 + 2 = 418_{10}$

Decimal (base-10): $642 = (6 \times 10^2) + (4 \times 10) + 2 = 600 + 40 + 2 = 642_{10}$

Hexadecimal (base-16): $3{:}13{:}6_{16} = (3 \times 16^2) + (13 \times 16) + 6 = 768 + 208 + 6 = 982_{10}$

Vigesimal (base-20): $3{:}13{:}6_{20} = (3 \times 20^2) + (13 \times 20) + 6 = 1{,}200 + 260 + 6 = 1{,}466_{10}$

Sexagesimal (base-60): $1{:}25{:}52_{60} = (1 \times 60^2) + (25 \times 60) + 52 = 3{,}600 + 1{,}500 + 52 = 5{,}152_{10}$

Note the notation conventions used here, which I shall use hereafter. When there is a possibility of ambiguity about the base of a number, a subscript defining the base follows the number. For example, the only way we can know whether 101 is a base-2 number, or a number of any other base, is either from the context in which it is used or by adding a subscript. For bases $b > 10$, colons separate positions, the same notation we use for time (hr:min:sec), and thus Hindu-Arabic numerals suffice and we do not have to define new symbols for base values greater than 10.

The difference between using base-10 and any other base is that our language is designed for base-10 use. We read the base-10 number 2,731 as two thousand seven hundred and thirty-one because we have defined the word thousand to mean 10^3, and the word hundred to mean 10^2. We do not usually have to use equation (1.3.1) explicitly to calculate the value of a base-10 number; rather, its use is implicit. With our base-10 language, we can only read other base numbers as a sequence of digits. For example, we read the base-2 number 1101 as one one zero one; we read the base-8 number 642 as six four two.

FUN QUESTION 1.3.1: What is the highest number you can count with a three-digit decimal number, and how many ways can you arrange three decimal digits?

ANSWER: 999 and 1,000. In any positional number system, the number of arrangements of symbols is always one more than the highest number counted because zero is one of the possible arrangements. The intuitive generalization of this result is that for a positional, base b, N-position number, b^N is the number of possible arrangements and $b^N - 1$ is the highest number that can be counted.

FUN QUESTION 1.3.2: A finger counter using positional base-2 counting (see FUN QUESTION 1.2.4) wants to count to one million. How many hands does he require?

Either from the previous tabulation of three-digit numbers for various bases, or from the ANSWER to FUN QUESTION 1.3.1, or simply intuitively, we can see that as the value of the base increases, the higher the number that can be counted with the samc number of positions. Thus, the larger the base, the more compact the writing tends to be.

There are also disadvantages to a large base. For facile use of a large base, we must memorize a large number of symbols, which can be a burden. However, the major disadvantage of a large base is that it can result in awkward arithmetic. To illustrate, let us consider addition. For easy pencil/paper calculation using the base-10 system, we memorize the addition table presented in table 1.3.3. As children, we required a few years to master this table. We also memorized two simple and obvious rules: $n + m = m + n$ and $0 + n = n$, where m and n are any numbers.

+	1	2	3	4	5	6	7	8	9
1	2	3	4	5	6	7	8	9	10
2		4	5	6	7	8	9	10	11
3			6	7	8	9	10	11	12
4				8	9	10	11	12	13
5					10	11	12	13	14
6						12	13	14	15
7							14	15	16
8								16	17
9									18

Table 1.3.3 *Base-10 addition table*

We can see from table 1.3.3 that the *arithmetic series*, $1 + 2 + 3 + 4 + 5 + 6 + 7 + 8 + 9 = 45$ (the number of shaded boxes), gives the number of entries it is necessary to memorize. For this series, the average term is $45/9 = 5$. In order to determine the burden in memorizing addition tables for other bases, and for other later use, it will be useful to generalize this numerical calculation as an algebraic equation. Let a series of N terms be $(a_1 + a_2 + \ldots a_N)$. If the average term of a series is a_{AVG}, then the sum, S, of a series is $S = Na_{AVG}$. This is the definition of the average term for any series.

An *arithmetic series* is one in which the difference between successive terms is constant. The series 1 + 2 + 3 + 4 + 5 + 6 + 7 + 8 + 9 is arithmetic, and the average term of an arithmetic series is clearly just the average of the first and last terms. Thus, for an arithmetic series we have a very simple equation for the sum,

$$S = N(a_1 + a_N)/2 \tag{1.3.2}$$

For a base-b addition table, $N = b - 1$, $a_1 = 1$, and $a_N = b - 1$, so that the number of entries is simply

$$S = (b - 1)b/2 \tag{1.3.3}$$

Table 1.3.4 summarizes the number of addition-table entries for various bases.

Base	2	8	10	16	20	60
Table entries	1	28	45	120	190	1,770

Table 1.3.4 *Addition-table entries*

Ten is probably the largest base for which addition can easily be done by memorizing an addition table. Does this mean that no base for which $b > 10$ can possibly be practical? No. It just means that addition by memorization of such a large number of table entries is not practical and some other method of addition must be used. Base-10 fortuitously has a base that is large enough to produce reasonably compact writing, but small enough so that it is possible to memorize an addition table, a multiplication table of comparable size, and symbols. (This is the most efficient way of doing pencil/paper arithmetic because recalling from memory is faster than a logical process.) While it evolved because we happen to have ten fingers, base-10 has survived because it adequately answers all requirements for everyday use.

Base-2 is not a candidate for everyday use because it lacks compactness. Current use is primarily in computer technology. Because current computer technology is based on arrays of ON/OFF switches, and an ON can be represented by a 1 and an OFF can be represented by a 0, we can appreciate that base-2 is currently the natural number system for digital computer programming. Base-2 numbers seldom have to be read by human eyes, and so what would be a confusing sequence of 1s and 0s for us, such as the nine-position base-2 number 110101011, is no problem for a computer's optical, electronic, or magnetic reading device.

Computer engineers must sometimes read base-2 numbers, and to minimize confusion they have invented the base-8 and base-16 number systems. Since a three-position base-2 number can count to seven ($111_2 = 7$), every three positions in a base-2 number can be replaced by one base-8 position. A computer can thus display the base-8 number 653 rather than the confusing base-2 number 110101011. Actually, transcription into base-16 is a practice that is more usual. Since a four-position base-2 number can be counted to fifteen ($1111_2 = 15$), every four positions in a base-2 number can be replaced by one base-16 position, resulting in a more compact display.

If it were worthwhile to choose the best possible number system for present-day use, my choice would be base-16. It produces a more compact number than base-10, but there are not too many symbols to be memorized. In current computer practice, the sixteen symbols are the ten Hindu-Arabic numerals plus the six letters A, B, C, D, E, F. To avoid confusion in everyday use, it would be necessary to replace these alphabetic symbols with new unique symbols. By design, base-16 is compatible with base-2 and hence with current computer technology, whereas conversion between base-10 and base-2 is somewhat of a nuisance (see FUN QUESTIONS 1.3.4 and 1.3.5). Base-16 addition and multiplication tables are uncomfortably large for memorization, but who cares in an era of ubiquitous electronic calculators.

I am not suggesting that there should be a conversion to base-16 for everyday use; this is just an academic exercise. The practical advantage of such a conversion pales before the chaos it would cause. Converting to base-16 for everyday use would be equivalent to converting to driving on the right side of the road in England.

FUN QUESTION 1.3.3: Convert the base-10 number 300 into a base-2 number.

ANSWER: To solve this you need to know the decimal values of base-2 positions:

2^9	2^8	2^7	2^6	2^5	2^4	2^3	2^2	2^1	2^0
512	256	128	64	32	16	8	4	2	1

Table FQ 1.3.3

Find the base-2 position with the largest value less than or equal to 300. This is position 8 with a value of 256. Subtract: 300 – 256 = 44. Find the base-2 position with the largest value less than or equal to 44. This is position 5 with a value of 32. Subtract: 44 – 32 = 12.

Find the base-2 position with the largest value less than or equal to 12. This is position 3 with a value of 8. Subtract: 12 – 8 = 4. This is equal to the value of position 2, and so the answer is 100101100.

Check the answer: 256 + 32 + 8 + 4 = 300.

This solution to FUN QUESTION 1.3.3 is by a sequence of mathematical operations. Such a step-by-step procedure is called an *algorithm* in modern math jargon. The word algorithm is a corruption of the name of the Persian mathematician al-Khowarizmi (circa 825), although the present meaning of the word algorithm is very different from the mathematics of al-Khowarizmi.

The algorithm just used is called, with what passes for humor among mathematicians, the *greedy algorithm*. Why greedy? Because we always subtract the base-2 position value that takes the biggest possible bite. Just as we have unknowingly been speaking prose since we first learned to speak, we have also unknowingly been using the greedy algorithm since we first learned to do base-10 division. For example, how do we do the division 73/27? First greedy step: multiply 27 by the biggest number (2 in this case) such that the product is less than or equal to 73, and subtract it from 73. And so on with further greedy steps. In succeeding chapters we shall encounter many other natural and intuitive applications of the greedy algorithm extending back to the very beginning of counting. In chapter 4, we shall see that this very intuitive *greedy algorithm* was a key element of Egyptian arithmetic. It plays a particularly important role in understanding Egyptian fractions.

Another common and contemporary use of the greedy algorithm is a way to calculate change dispensed in a cash transaction. Rather than considering how we do such a calculation, which may not use the greedy algorithm, consider the algorithm that the "brain" of a cash register in the United States uses to choose which coins come sliding down the chute of an automatic coin dispenser. Consider the calculation of change when purchasing a 9¢ item with a $1 bill. The greedy algorithm dispenses change with the minimum number of coins.

- 100 – 9 = 91. What is the greediest coin equal to or less than 91¢? 50¢.
- 91 – 50 = 41. What is the greediest coin equal to or less than 41¢? 25¢.
- 41 – 25 = 16. What is the greediest coin equal to or less than 16¢? 10¢.
- 16 – 10 = 6. What is the greediest coin equal to or less than 6¢? 5¢.
- 6 – 5 = 1. What is the greediest coin equal to or less than 1¢? 1¢.
- 1 – 1 = 0. Change complete.

This transaction can be between a purchaser and an automatic vending machine. The only arithmetic required of the purchaser is to know that $1 is greater than 9¢. Even if a salesperson is involved, no ability to do arithmetic is required: at most, the salesperson must key in the cash received, a laser scanner reads a bar code (a binary code) and enters the price directly into the cash register, whose "brain" then does all of the necessary arithmetic. This transaction nicely illustrates recent evolution in arithmetic. A small cadre of mathematically competent people converts arithmetic into computer programs that relieve the vast majority from ever having to perform more than the most trivial arithmetic. For most people, most of the arithmetic once so boringly and painfully learned will soon atrophy into a vague memory.

The conquests of Islam in the eighth century led to the acquisition and translation into Arabic of Greek and Hindu mathematical texts. In this manner such texts were preserved and eventually translated into Latin and became available in Europe. The writings of al-Khowarizmi were the most important. He used the Arabic word *al-jabr* to define the rules for transferring a term from one side of an equation to the other side. For example, in *algebraic notation*, $a + b = c$ can also be written as $a = c - b$. Eventually the Latinized *al-jabar*, now *algebra*, came to stand for all the rules of arithmetic using *algebraic notation*.

FUN QUESTION 1.3.4: Compose the base-2 addition table and add the base-2 numbers 11010100 and 10101111. Give the answer as base-2, base-8, base-16, and base-10 numbers.

FUN QUESTION 1.3.5: Compose the base-8 addition table and add $234_8 + 456_8$. Give the answer as base-8, base-16, base-2, and base-10 numbers.

FUN QUESTION 1.3.6: Zorbi, an alien from the planet Geek of the star Gamma Centuri in the Andromeda galaxy, landed in my backyard, of all places. We eventually learned how to communicate, and he told me that his spaceship had a crew of 2,431 men. However, on a visit to his spaceship I counted a crew of only 366. Had he lied to me, and did this mean that he was really not friendly? I could not believe it; he seemed like such a nice guy. The fate of Earth depended on my resolving this discrepancy. Then it occurred to me that perhaps he was simply not counting in base-10. Was there a base-*b* such that $2{,}431_b = 366_{10}$? Use the greedy algorithm.

FUN QUESTION 1.3.7: In 1952 Emil Zatopek won three Olympic Gold Medals in long-distance races: 5,000 meters, 10,000 meters, and marathon. His time for the marathon was 2:23:3, which means 2 hours, 23 minutes, 3 seconds. Express this time in seconds.

ANSWER: One hour equals sixty minutes and one minute equals sixty seconds, so Zatopek's time in seconds can be written as $(2 \times 60^2) + (23 \times 60) + 3 = 8{,}583$ seconds. Zatopek's feat was athletically remarkable, but what is mathematically remarkable is that 2:23:3 is a base-60, or sexagesimal, number. It is a vestige of the number system used by the Babylonians, whose empire ceased to exist some 2,500 years ago. In section 3.2 we shall see why the Babylonians, presumably a people with ten fingers like most of us, used a sexagesimal system.

Zatopek's time for 5,000 meters was 14:06.6, which means 14 minutes, 6 seconds, and 6 tenths of a second. For times less than seconds, the measurement reverts to the decimal system and ends up as a sexagesimal-decimal, mixed-units system. Such mixed units would be misleading to someone who is not a sports fan and is not aware of this almost-unmarked, units-change convention. In section 3.3, we shall see how archeologists, who did not appreciate an unmarked units change, have incorrectly transcribed Mayan numbers.

2

THE BIRTH OF ARITHMETIC

2.1 Pattern Recognition Evolves into Counting

Counting is the simplest and most used arithmetic operation and is therefore the first operation invented by humans. How long have humans been counting: ten thousand years, a hundred thousand years, a million years? The answer depends on just how counting is defined. The appreciation of *quantity* as a property of something is instinctive for most animals. The essential survival instinct of whether to *fight or flee* is, among other things, based on instinctive, quantitative appreciation of *too big* and *too many*. However, this should not be considered counting, but only *pattern recognition*. When did the brain of a modern human's hominid antecedents (a hominid is an upright-walking primate) make the transition from just pattern recognition to counting with symbols such as fingers or scratches or vocalized names for numbers?

The interpretation of recent fossil finds suggests that hominid evolution began when some chimpanzee-like creatures began walking upright at least seven million years ago. Weather change in Africa had converted vast stretches of jungle into grasslands (savannas), and some chimplike creatures began to forage for food on the savannas. Those with a genetic variation that resulted in better ability to stand and run upright had a survival advantage. The evolving bipedalism left hands free for using tools and weapons and carrying food, which also conferred a survival advantage. A bigger brain enabled better tool and weapon making and exploitation, and thus began the evolutionary path that eventually produced modern humankind.

The first tools and weapons were just rocks, bones, and sticks with minimal fabrication, and no evidence remains of such artifacts. Nonetheless, we are certain of such tool use because tool use by chimpanzees in the wild has now been observed. However, it took about five million years for hominids to evolve into *Homo habilis*, who fabricated tools and weapons from stone, which have survived and been identifiable as tools and weapons. Figure 2.1.1 schematically illustrates the growth in brain size. Growth attributable primarily to the hands-brain feedback cycle occurred in the first five million years of hominid evolution. Starting with the brain size of a chimpanzee today as a reasonable approximation for the brain size of the first hominids, it took about five million years for the brain size to increase from about 350 to 500 cc. Although the evolution of bipedalism enabled the hands-brain feedback cycle, bipedalism per se does not require a larger brain since it is primarily just evolution of the skeleton. Interesting confirmation of this is in recent developments in robotics, which show that two-legged robots require even less "brains" than four-legged robots.

Starting around three million years ago, hominid brain size began to grow much more rapidly, almost tripling in size in some three million years to the size of the human brain today, which averages about 1,350 cc. This increased rapid growth is presumably due primarily to the evolution of language in the throat-brain feedback cycle. Thus, within the last three million years, grunts and screeches evolved into words, and words evolved into combinations of words, and combinations of words evolved into concepts, and pattern recognition evolved into real counting. Incidentally, figure 2.1.1 shows that the classically invoked explanation of hominid brain growth, the hands-brain feedback cycle, is apparently much less important than the throat-brain feedback cycle.

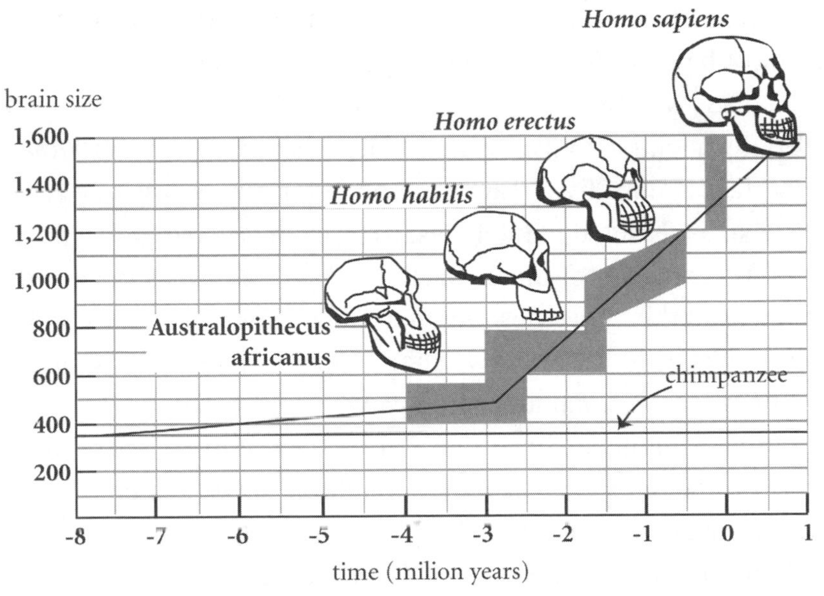

Figure 2.1.1 *Brain-size evolution in hominids*

Certainly better communication enabled hunting and gathering that was more efficient, which enhanced chances that a clan of early hominids would survive, but it does not explain why individuals in the clan with a genetic variation that resulted in better language skills preferentially survived. The reason is presumably that language skill is to humans, what the tail is to the peacock. Any modern women who has ever been "sweet talked" into having sex is probably following a predilection that is possibly some three million years old.

The critical discovery that enables distinguishing between hands-brain feedback and throat-brain feedback contributions to brain growth is the recent fossil evidence that shifts the beginning of bipedalism backward in time by some four million years. In drawing figure 2.1.1, I have arbitrarily chosen eight million years ago for this beginning; it must be before the first fossil evidence, but how much before cannot be known. I have also drawn figure 2.1.1 with a sharp break at three million years ago just to make the onset of the throat-brain feedback cycle more apparent. In reality, this must have been a gradual change.

The growth of the "brain" in my legless/armless computer called my attention to the separate role of language in brain growth. In the last twenty-five years my computer's memory has grown from a size measured in kilobytes, to a size measured in megabytes (1,000 times larger, or more exactly, $2^{10} = 1{,}024$ times larger), to a size measured in gigabytes (another 1,000 times larger), as its language capabilities have grown more sophisticated as it duplicates anthropomorphic communication capabilities. (A byte is an eight-digit binary number. Since there are $2^8 = 256$ different arrangements of one byte, each keyboard symbol can be defined by a different arrangement of one byte.)

From figure 2.1.1 we see that *Homo sapiens* brain size is probably destined to continue increasing, although it will take thousands of years before such a trend can be determined because of the variation in brain size from individual to individual. However, it is not clear whether this extrapolated increase will require a larger cranium or this increase will be contained in external computers that will relieve the human brain of some of its tasks—a process that is already well under way.

It was not until about one hundred thousand years ago that *Homo sapiens* began to leave ritualized-burial evidence of contemplative thinking. By about thirty thousand years ago, they also began to leave cave paintings, further evidence of contemplative thinking and a culture-wide belief system. A culture-wide belief system requires intelligence and good communication, so this is the first evidence that humans were sufficiently intelligent to have the concept of number and sufficiently articulate to have words for some numbers. Thus, real counting with names for numbers almost certainly began before about 30,000 BCE. But lack of evidence is not evidence of lack, and real counting could have possibly begun as long ago as one millions years, when brain size grew to about 900 cc, which is the minimum size some experts set for normal humanlike thinking. I shall somewhat arbitrarily choose the era with unambiguous physical evidence of contemplative thinking, roughly 50,000 BCE, to define the beginning of real counting and the birth date of arithmetic.

Hardly any interpretation of an anthropological finding about early hominids has not been challenged or changed. I will not be surprised if my estimate for a birth date of arithmetic will also be challenged or changed. The choice of 50,000 BCE as a time by which counting began suffices to show that mathematics made little progress for at least some tens of thousands of years of hunter-gatherer culture. Primitive hunter-gatherers simply had little need for arithmetic. Better understanding of brain function, rather than new anthropological findings, may be able to move the birth date of arithmetic to a significantly earlier era—and I expect it will.

2.2 Counting in Hunter-Gatherer Cultures

At the end of the nineteenth century, the Cambridge Anthropological Expedition to Oceania recorded an interesting variation on finger counting. On islands in the Torres Strait between New Guinea and Australia, Stone Age natives included other body parts to define numbers, as illustrated in figure 2.2.1.

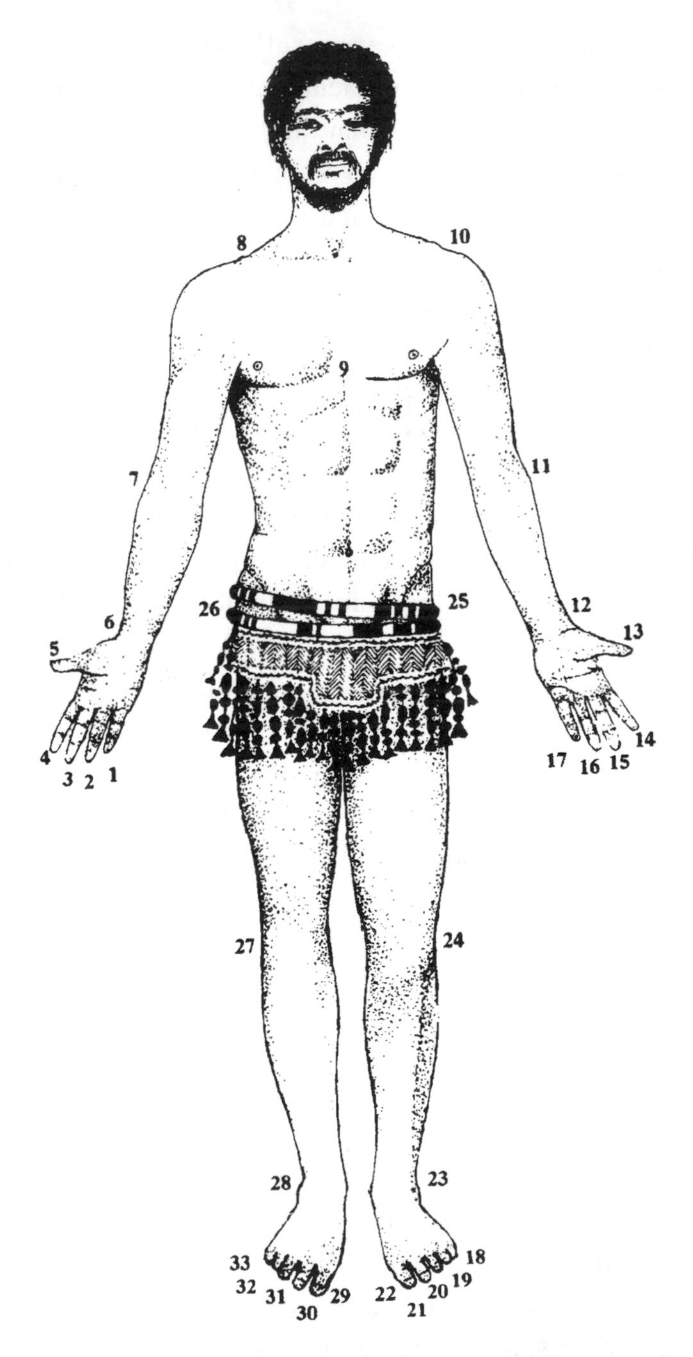

Figure 2.2.1 *Body-parts counting*

They communicated a number by pointing to the appropriate body part or by vocally naming a body part. Presumably, millennia ago they started counting on their fingers, then added toes, and as a need arose for even higher counts, they added other body parts. This system employs the concept of position to define thirty-three names for numbers, but this does not make it a *positional* system because it lacks the essential concept of *replacement*. For easy counting and communication with this number system, it is necessary to memorize the values assigned to thirty-three body parts—not very easy, especially for larger numbers that are infrequently required.

The awkwardness of this body-parts system is even more apparent when considering its use in arithmetic. This implies that at least this particular Stone Age tribe did not do much if any arithmetic beyond simple counting. Facile addition with a number system that uses memorized symbols or names requires memorizing an addition table, just as we presently do in pencil/paper, base-10 addition with ten memorized symbols, 0 . . . 9. Employing equation (1.3.2), we can easily calculate that this body-parts counting would require memorizing an addition table with 561 (= 33 × 34/2) entries, rather beyond the capabilities of most people. These Stone Age people certainly never did addition using a memorized addition table with all of the numbers defined by body parts, although they may have memorized a few frequently arising simple additions such as *right wrist* plus *right elbow* equals *left thumb.*

FUN QUESTION 2.2.1: Using Torres Strait body-parts counting, how would you express the sum, 27 + 11, in terms of body parts?

It is interesting to compare this body-parts counting to the finger-counting system described in section 1.2, where all fingers on the right hand have a value of one and all fingers on the left hand have a value of five. This simple system could count to thirty, almost as high as with Torres Strait body-parts counting, and with minimal memorization. Without *replacement*, the number of symbols quickly becomes excessive and the arithmetic becomes awkward.

* * * * *

FUN QUESTION 1.2.4 illustrated that the base-2 number system does not require sophisticated mathematics to derive. The derivation was intuitive, easy, and only required hands—humankind's first calculating machine. Base-2 counting could have been invented in the primitive, *hunter-gatherer* stage of human evolution, and some scholars believe it was. But was it?

Into the nineteenth, and even into the twentieth, century, a few primitive hunter-gatherer cultures remained untainted by contact with surrounding civilizations. It is from studies of the present-day language of these cultures that we also have some appreciation of the number systems of ancient hunter-gatherers. In the last fifty thousand years, these hunter-gatherers may have improved their counting systems and their ability to articulate so we cannot know exactly how they counted some fifty thousand years ago. But we can be reasonably sure that their counting was not more advanced than that of present-day hunter-gatherers.

A typical example of present-day (actually early twentieth-century), primitive, hunter-gatherer counting is that of the Aborigines of Australia, although I am not sure if any Australian Aborigines remain who have not now learned and adopted decimal counting and English nomenclature. They seldom count beyond six and identify numbers as combinations of just the words *one* and *two*. Table 2.2.1 interprets such counting as simple addition of 1s and 2s. A 1-for-2 replacement shortens the number of words required to express a number.

Base-10 number	Aborigine language	Additive interpretation
1	one	1
2	two	2
3	two one	2+1
4	two two	2+2
5	two two one	2+2+1
6	two two two	2+2+2

Table 2.2.1 *Additive interpretation of two-word Australian Aborigine counting*

However, if we interpret twotwo as a new word with a value of four, then table 2.2.2 shows how this essentially makes the number system into a three-word system and interprets the Aborigine counting as consistent with base-2 and in accord with the general definition of a number system with a base, equation (1.3.1).

Base-10 number	Aborigine language	Base-2 interpretation
1	one	1
2	two	2
3	two one	2+1
4	twotwo	4
5	twotwo one	4+1
6	twotwo two	4+2

Table 2.2.2 *Base-2 interpretation of two-word Australian Aborigine counting*

Although with this imaginative stretch it is possible to interpret this Aborigine counting as base-2, in reality it surely is two-word additive and not base-2. The slightly different language used by a different clan proves this. Their three-word counting exhibited in table 2.2.3 uses the words *one*, *two*, and *three*, and the only possible interpretation is that the counting is additive.

Base-10 number	Aborigine language	Additive interpretation
1	one	1
2	two	2
3	three	3
4	three one	3+1
5	three two	3+2
6	three three	3+3

Table 2.2.3 *Additive interpretation of three-word Australian Aborigine counting*

FUN QUESTION 2.2.2: Assuming that only the words *one*, *two*, and *three*, or combinations of them, are used to count in a base-3 system, express decimal numbers from 1 to 10 in base-3.

FUN QUESTION 2.2.3: Write the decimal numbers 1 to 10 as positional base-3 numbers using the three symbols 0, 1, 2.

The fact that in two-word Aborigine counting the word *two* appears repeatedly and in three-word Aborigine counting the word *three* appears repeatedly has been interpreted by some mathematically naïve anthropologists as evidence for base-2 and base-3 counting.

A better way to account for such Aborigine counting languages is as simply avoiding the need to define new words for seldom-used numbers (numbers greater than two or three). In fact, this Aborigine language for numbers is an excellent example of a linguistic principle, which states that frequently used words are memorized while seldom-used words are based on rules. This can be illustrated by present-day use of irregular and regular verbs. Table 2.2.4 compares the use of *be*, the most used verb, with the use of *factor*, a rarely used verb (except in mathematics).

If hunter-gatherers seldom counted to six, it would be possible for a whole generation never to hear *six*, and thus the word would

verb	present tense	past tense
irregular verb, frequently used, tense change by memory	am	was
	are	were
	is	was
regular verb, seldom used, tense change by rule: add **ed**	factor	factor**ed**

Table 2.2.4 *Example of linguistic principle*

not survive. For the two-word counting just considered, the frequently used words *one* and *two* would be remembered, and the examples from frequent use of *two one* (rather than a new word *three*) and *two two* (rather than a new word *four*) would be sufficient to define (intuitively if not consciously) two simple rules:

1. Add words *one* and *two* to define numbers greater than two.
2. Use the maximum number of *twos*. (Recognize the *greedy algorithm*? See the discussion following FUN QUESTION 1.3.3.)

With these rules, any number, however large, can be defined. Of course, for large numbers this leads to very cumbersome language. So what? They seldom needed large numbers.

In addition to these examples of the Australian Aborigine language, studies done in the early twentieth century of Native North American languages concluded that some 30 percent of the tribes used base-2. However, none of these old conclusions had the advantage of recent linguistic *words-and-rules* understanding about how the brain processes language. These old studies, which concluded that a large fraction of hunter-gatherer counting was base-2, are of dubious validity. My guess is that Stone Age hunter-gatherers never counted with base-2, or any other base. A base is defined when a number system uses a sequence of replacements with the same replacement number. A base can only be defined by counting to high numbers, so it is unreasonable to conclude that just a few, small integers can define a base. However, recent anthropological studies of Native American number systems have more insightfully defined such systems as 2-based rather than base-2.

Another example of mathematically naïve anthropologists misinterpreting primitive counting is the concept of *concrete numbers* versus *abstract numbers*. At the end of the nineteenth century, Franz Boas (1858–1942) studied Native American tribes on the west coast of Canada. He concluded that they used different sets of numbers to count men, canoes, long objects, flat objects, round objects, and other objects. This is *concrete* counting, also called *archaic* counting. The assumption was that such usage showed lack of understanding of the concept of an *abstract number*. The perception was that the natives did not understand that saying *two* canoes and *two* men expressed the shared property of *twoness*. But they did understand that saying *good* canoes and *good* men expressed the shared property of *goodness*. This makes little sense.

Rather than representing concrete counting, I think that Native American counting is better explained as a systematic use of a syntactic device still used in English and in many languages: the addition of a suffix to a number word for more concise wording. Examples in English are one-sided argument, two-faced women, tricolor flag, and fourfold profit.

Most Native American languages had words for numbers based on names for fingers. A generic rendering in English of such practice would be *pinky*, *next*, *middle*, *pointer*, and *hand* for the numbers 1, 2, 3, 4, and 5. This is clearly using a one-for-one correspondence between quantity of fingers and quantity of objects. This is itself proof of an abstract concept of number. However, when making a quantitative statement they added a suffix that usually added a geometric property for a more complete mathematical description in one word. For example, in my generic Native American language, two basketballs would be *nextround* basketballs, three pancakes would be *middleflat* pancakes, and four strings would be *pointerlong* strings. However, for complex shapes like a canoe, the suffix was the word *canoe* itself. Thus, a man with five canoes would be concisely described as a *handcanoe* man, just as in English we might call him a five-canoe man.

However, there is a difference between modern use of such syntax and Native American use. We only use such syntax occasionally, while the Native Americans, who used such syntax, always used it. My guess is that this is because Native American dual use of names of fingers creates some ambiguity about whether the name refers to a finger or to a number. The addition of the suffix when a number is implied removes the ambiguity. Words in modern languages for the first ten digits are possibly also descended from ancient names for fingers, but any such meaning has long been forgotten.

The concept that there was a transition from concrete counting to abstract counting is superfluous.

The propagation of this concept is largely due to the statement by the prestigious mathematician Bertrand Russell (1872–1970): "it . . . required many ages to discover that a brace of pheasants and a couple of days were both instances of the number 2." However, Russell had never personally studied the language of primitive peoples, as Boas and other anthropologists had. He was simply repeating their apparently well thought-out and superficially reasonable conclusions. Had he bothered to consider the etymology of these words, he would have realized that the English word *couple* comes from the Latin *copula*, a bond; it takes two to tango and two to make a bond. The English word *brace* comes from the Latin word *bracchia*, meaning arms, from which are descended such English words as *embrace* and *braces* (what Americans call suspenders). These words probably entered English usage from French. We can easily envisage how two pheasants tied together at their feet were easier to carry. Thus, two braced pheasants came to be called a brace of pheasants. Russell mistakenly interpreted metaphoric language as concrete counting. "Problems arise when we begin to believe literally in our own metaphors" (quoted from *The Da Vinci Code*, by Dan Brown, 2002).

Mathematical competence generally ascribable to Stone Age hunter-gatherer cultures is:

1. Counting extends only to small numbers, usually to less than ten.
2. The concept of a number base is neither understood nor used.
3. The concept of addition is understood.
4. Number is understood as an abstract concept.
5. At least some Stone Age hunter-gatherers invented and used the replacement concept.

* * * * *

Surprisingly, there is a much-heralded archeological find of supposed mathematical significance dated from about twenty thousand years ago, when humans were still primitive hunter-gatherers. A geologist, J. de Heinzelin, found the so-called Ishango Bone in 1960, in central equatorial Africa. Figure 2.2.2 is a sketch of this artifact.

The scratches were apparently intentionally made and are not just tooth marks by either man or beast. Prevalent opinion is that the scratches on the bone represent some kind of tally. But that is just a guess because the scratches could have been purposefully made to provide a better grip on a tool. The bone is indeed the handle of a tool with a piece of quartz attached to one end. The markings may also simply be the product of the Stone Age equivalent of whittling, or they may have been made for some other non-mathematical reason. Despite these obvious reservations, this one

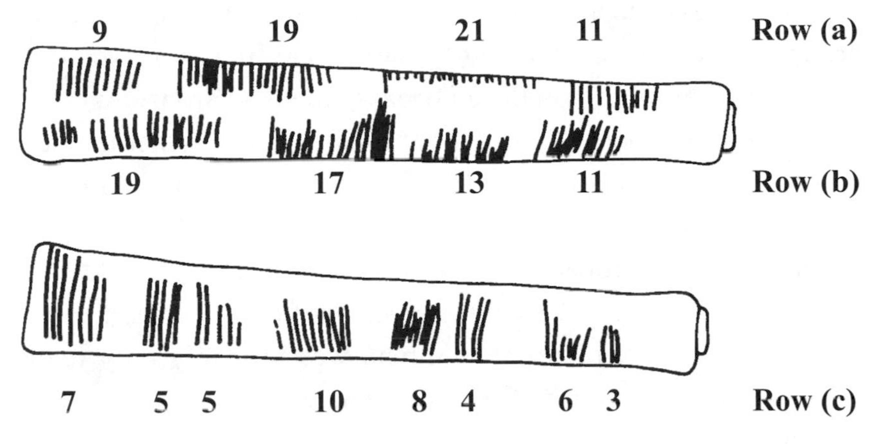

Figure 2.2.2 *Ishango Bone*

small bone has inspired many new speculations about Stone Age hunter-gatherer mathematical competence.

The numbers in Row (a) have been taken as evidence for base-10 understanding because all of the scratches can be accounted for as 10 – 1, 20 – 1, 20 + 1, and 10 + 1, a rather imaginative way of concluding that the numbers 9, 19, 21, and 11 imply base-10 usage. It was even proposed that, since central equatorial Africa is up-river from Egypt, this is evidence that the eventual adoption of base-10 by the Egyptians was by *diffusion* of this use from central Africa.

At the beginning of the twentieth century, diffusion was the generally accepted theory for why different cultures used the same inventions. It is now recognized that *independent invention* also frequently occurs when similar conditions exist in different cultures. The condition required to choose base-10 is simply the universal existence of ten fingers on humans. However, distinguishing between *diffusion* and *independent invention* is a problem that we shall encounter repeatedly, and it is not always easy to decide which mechanism has operated.

The numbers in Row (b) are the prime numbers between 10 and 20, which has been taken as evidence that there was already appreciation of the concept of prime numbers. If an archeologist found a bronze sword and dated it from about 20,000 BCE, this would be viewed with extreme skepticism because the first Bronze Age artifacts date from about 3500 BCE. Mathematics and other intellectual concepts also evolve in a sequence that can be approximately dated. The concept of division, which must precede the concept of a prime number, probably did not evolve until after 10,000 BCE (see sections 4.2 and 5.2) and the emergence of herder-farmer cultures. The concept of prime numbers was probably only really understood after about 500 BCE by Greek mathematicians (see section 6.2).

The numbers in Row (c) include 3 and 6, 4 and 8, and 5 and 10, among others, which has been interpreted as exhibiting multiplication by two. However, no attempt has even been made to explain why a tally of something should exhibit multiples of two, prime numbers between 10 and 20, and some numbers that are almost multiples of ten. I have no objection to the conclusion that the markings do represent a tally, even though the evidence is not particularly convincing. If a man can count on his fingers and vocalize names for numbers, he can also represent the numbers by scratches in order to make a permanent record. Note that if the scratches are a tally, it is abstract counting without association with any particular object.

Every discoverer of an archeological artifact wants it to be the find of the century. Thus, we can understand that the discoverers and interpreters of the Ishango Bone tried to milk their artifact for more than it was worth. However, without mathematical perspective, they did not realize that concepts such as a number system with a base, prime numbers, and even the multiplication operation, which seem so elementary to people with a modern education, were not understood twenty thousand years ago.

Ethnomathematicians, who commendably seek to instill pride

about their African heritage in black Americans, have made the Ishango Bone into a symbol of precocious African mathematics. This is an unfortunate choice because of its dubious interpretation. Other artifacts of comparable age, less ambiguously marked as tallies, have been found in Europe and Africa. *Homo sapiens*, whether in Africa or Europe, in a Stone Age hunter-gatherer stage learned to express his quantitative instinct as discrete number symbols: fingers, scratches, and sounds. When his skin evolved to a lighter hue in the shaded forests of Europe, *Homo sapiens* did not also lose his ability to count. Apparently, skin color has little to do with mathematical ability.

3

PEBBLE COUNTING EVOLVES INTO WRITTEN NUMBERS

3.1 Herder-Farmer and Urban Cultures in the Valley of the Nile

Before the tenth millennium BCE, Stone Age technology was similar in hunter-gatherer cultures from Europe to Asia to the Americas. The fact that there were not large differences means that knowledge had diffused worldwide. Diffusion occurred by some combination of migration and contact. If the average diffusion rate were at least about 20 km per year, which appears to be a reasonable assumption for a mobile hunter-gatherer society, it would have taken less than about one thousand years for innovations to spread halfway around the world, sufficiently rapid to account for similar skills throughout a slowly evolving world.

Then significant differentiation began. Herding and farming originated on the fertile plains of the great river systems: the Nile in Egypt, the Tigris and Euphrates (Mesopotamia) in what is today Iraq, the Indus in what is today Pakistan, and the Yellow and Yangtze in China. This Stone Age transition from hunter-gather to herder-farmer cultures is known as the *Neolithic Revolution*. Primitive herder-farmers must keep track of more material possessions than hunter-gatherers must. They may also produce surplus, which they can trade. In addition, they must pay taxes and count the passage of days to track the seasons. Motivated by a need for a better counting system, pebble counting was the solution.

We know that pebble counting was used by primitive herder-farmers because such cultures still exist today that use this system. As an example, I quote a lyrical passage from an autobiographical novel, *An American Visitor*, by Joyce Cary, about the British Empire in Africa at the beginning of the twentieth century. "In another month or six weeks the first rains would fall and men would begin to recon time again, dropping a stone in a calabash for each day until the due date for corn and grain to be ripe." This quote also illustrates primitive calendar calculation. There is no need to count all the days of the year, rather only the relatively few days numbered from some event of critical importance to a herder-farmer, such as the beginning of a rainy season.

There have also been many present-day observations of the 1-for-10 replacement method of pebble counting in primitive cultures as previously depicted in figure 1.2.1. Since possibly starting with pebble counting in Egypt as early as ten thousand years ago, base-10 is now the number system of choice throughout the world.

The fourth millennium BCE was a time of great change: oxen were first hitched to a plow, irrigation was first used, the wheel was invented, ceramic technology matured, and the use of copper and bronze tools began. With improved agriculture producing a surplus, it became possible to feed an urban population of merchants, artisans, soldiers, priests, and, of course, tax collectors; cities began to grow. Now there was a need for more efficient record keeping, and the solution was written numbers and language. By around 3000 BCE, written records were used in Egypt and Mesopotamia.

It is generally accepted that cultures in Egypt and Mesopotamia were the first to develop written mathematics. I am not sure if this is true, or if it is simply a consequence of accidents of climate, choice of writing materials, and history, so that only here have relevant documents survived, been found, and been decipherable.

Decimal number	**1**	**10**	**100**	**1,000**	7,432
Heiroglyphic symbol	𓏺	𓎆	𓍢	𓆼	𓆼𓆼𓆼𓆼𓆼𓆼𓆼 𓍢𓍢𓍢𓍢 𓎆𓎆𓎆 𓏺𓏺

Table 3.1.1 *Egyptian hieroglyphic number symbols*

Egyptian written numbers replicated the 1-for-10 replacement scheme of additive pebble counting by simply substituting written symbols for pebbles. Table 3.1.1 exhibits some symbols of the Egyptian hieroglyphic number system. Reading Egyptian hieroglyphic numbers is easy and obvious for us who are familiar with the base-10 system. Since it is an additive system, the symbols can be arranged in ascending order from left to right or from right to left or vertically or even in random order and the value remains the same.

These symbols possibly represent a finger (1), a heel bone (10), a rope coil (100), and a lotus plant (1,000). Why they chose these particular symbols is not known, but we can be sure they had some mystic significance in their symbol-conscious culture. There were other symbols for higher powers of ten so that any number, however large, could be written just by adding new symbols. There is no limit to the number of hieroglyphic symbols required, but use of numbers higher than one hundred thousand was rare, so in reality values of only about six symbols had to be memorized.

Addition and subtraction operations are also easy with hieroglyphic numbers. There is only one rule: replace ten symbols with one symbol of ten times the value. Even after the invention of writing, Egyptians continued to use pebbles to add and subtract (see table 1.2.1 for base-10 abacus addition). The Romans still used pebbles, which they called *calculi* and from which are derived the English words *calculation* and *calculus*. The English called pebbles *counters*, and the table on which business was calculated acquired the name *counter*; and so it is still called even though an electronic cash register has replaced the pebbles.

FUN QUESTION 3.1.1: Convert 456 and 567 into hieroglyphic symbols and add them.

Hieroglyphic writing was used mainly for inscriptions on monuments and walls of palaces, temples, and tombs. For writing on papyrus, which was used for most administrative and literary documents, the ancient Egyptians used a cursive script called *hieratic*. Although derived from hieroglyphic symbols, the hieratic symbols gradually evolved into symbols that bore little resemblance to their hieroglyphic origins. In addition to making writing easier, new symbols replaced repetitions of the same symbol, which made written numbers more compact. Table 3.1.2 presents one variation of hieratic number symbols, which varied with time and from scribe to scribe.

To write numbers in hieratic script requires memorizing many more symbols. As can been seen in table 3.1.2, to write numbers up to 9,999 requires memorizing the values of $4 \times 9 = 36$ symbols, but the changes in many symbols as the repetition number changes are reasonably obvious, so such memorization was probably not very burdensome.

FUN QUESTION 3.1.2: Convert 456 and 567 into hieratic symbols and add them.

1	2	3	4	5
6	7	8	9	10
10	20	30	40	50
60	70	80	90	100
100	200	300	400	500
600	700	800	900	1,000

Table 3.1.2 Hieratic number symbols

Nowadays base-10, pencil/paper arithmetic requires memorization of an addition table with forty-five entries (see table 1.3.3). If Egyptian scribes used the addition-table method, hieratic, pencil/paper (brush/papyrus would be a better description) addition with numbers up to 9,999 would require four tables, each of forty-five entries: one for units symbols, one for tens symbols, one for hundreds symbols, and one for thousands symbols. All four tables are somewhat similar, so it is not as difficult as memorizing $4 \times 45 = 180$ completely independent entries. However, add to that the required memorization of thirty-six hieratic symbols, and it is quite a memorization burden.

With this memorization burden, if the addition-table method of addition was used, Egyptian scribes would have required hieratic addition tables, at least during a learning stage, but not a single hieratic addition table has ever been found. It is interesting to speculate why. One possibility is that such tables were used, but simply none has survived. Thousands of Egyptian scribes labored at calculations for thousands of years, and therefore they must have written thousands upon thousands of hieratic, mathematical papyri. Only a few papyri out of these thousands have survived, so the probability of finding addition tables, even if they once existed, is not great. The other possibility is that they did not use the addition-table method. They surely also did abacus addition and subtraction that obviated the need for addition tables. Another possibility is that they calculated on a scratch pad using hieroglyphic symbols, which is essentially just abacus arithmetic with written symbols. As we shall see in succeeding sections, we know something about how the Egyptians did the more sophisticated operations of multiplication and division, but we still are not sure how they did hieratic addition and subtraction.

Quantitative measurement, in addition to counting of things, became a requirement after about 10,000 BCE. This imposed new demands on number systems. Consider the following scenario: a woman is making a garment and invents the natural, intuitive measuring system of using the widths of her fingers. She measures something as 19 finger widths, which she may record as 19 pebbles. If she knows the 1-for-10 pebble replacement system, she may conceptualize and even record her measurement as 1 large pebble and 9 small pebbles. However, there is another natural, intuitive way of conceptualizing and recording her measurements. In practice, measurement in units of finger widths is by a series of handbreadth-to-handbreadth placements, so she defines a handbreadth as a natural, intuitive, new unit, with 1-for-4 replacements of hands for fingers. Now, letting a larger pebble represent a handbreadth, she conceptualizes and records her measurement as 4 large pebbles and 3 small pebbles. Our prehistoric seamstress has now invented the use of *mixed units* of handbreadths and fingers.

The 1-for-10 replacement scheme that was so natural and intuitive for counting is no longer obviously the better system for units of measurement. What to do?

- Solution 1: Retain the natural 1-for-10 replacements for counting, and retain the various natural replacements for measuring. This method largely accounts for the English system, still used in the United States although not officially in England since 1965. But anybody born in England before about 1955 surely still thinks in "English" rather than in "metric." It is difficult to abandon either the language or the measures we learned as children.
- Solution 2: Define new, not-so-natural units of measurement so that 1-for-10 replacements also define measurement units. This is an important component of the present, almost universally adopted metric system, and was an important part of the Egyptian solution. It might be an overstatement to say that the Egyptians realized more than five thousand years ago what the rest of the world is only now realizing. They possibly simply did what came naturally and used the same intuitive 1-for-10 replacement system for measuring units as they did for counting. It was fortuitously a very good choice.
- Solution 3: Modify the natural 1-for-10 replacements for counting by adopting some of the natural replacements for measurements. This appears to have been the Babylonian solution (see section 3.2). It was fortuitously a bad choice (see section 5.1).

The Egyptian definitions of land areas illustrate their use of base-10 measuring units:

$$tA = 10 \times 10 \text{ (royal cubits)}^2 \cong 27.5 \text{ m}^2$$
$$xA = 10 \times 100 \text{ (royal cubits)}^2 \cong 275 \text{ m}^2$$
$$sTAt = 100 \times 100 \text{ (royal cubits)}^2 \cong 2{,}750 \text{ m}^2$$

Over the many millennia of Egypt's existence, unit names and definitions changed. This tabulation relates to the Old Kingdom (Dynasties III–VI) in the third millennium BCE. The Great Pyramids of Giza were built in this era.

Figure 3.1.1 shows that the Egyptians perceived the construction of areas one dimension at a time. This is not so very different from how we now conceptualize geometry: the locus of a mathematical point generates a line; the locus of a line generates a plane; the locus of a plane generates a solid. The ancient Egyptians did not have the abstract, mathematical concept of an infinitesimally small, dimensionless, mathematical point. But starting with a finite square cubit, they generated area and volume by the same process.

When the counting system and the measuring system use the same base, recording in better-visualized *mixed units* is easy. Thus, an area of 12,345 square royal cubits could have been expressed in mixed units as 1 *sTAt*, 2 *xA*, 3 *tA*, and 45 square royal cubits. This would have been confusing because there are too many different units. It could also have been expressed as 12 *xA*, 345 square royal cubits. The rule for normal practice today is, and probably always was, not to use more than two different units to define a measurement. Violation of this intuitively understood rule can be comic: "How long is it since you had your last cigarette?" she asks. "Three months, two weeks, four days, five hours, seventeen minutes," and with a glance at his watch, "and thirteen seconds," he answers.

Other units that were multiples of two and ten of the *sTAt* also defined land areas. In English units, the *sTAt* is about 0.7 acre. We shall see in section 3.2 that multiples of a unit of about the size of an acre defined agricultural areas in many cultures in many eras.

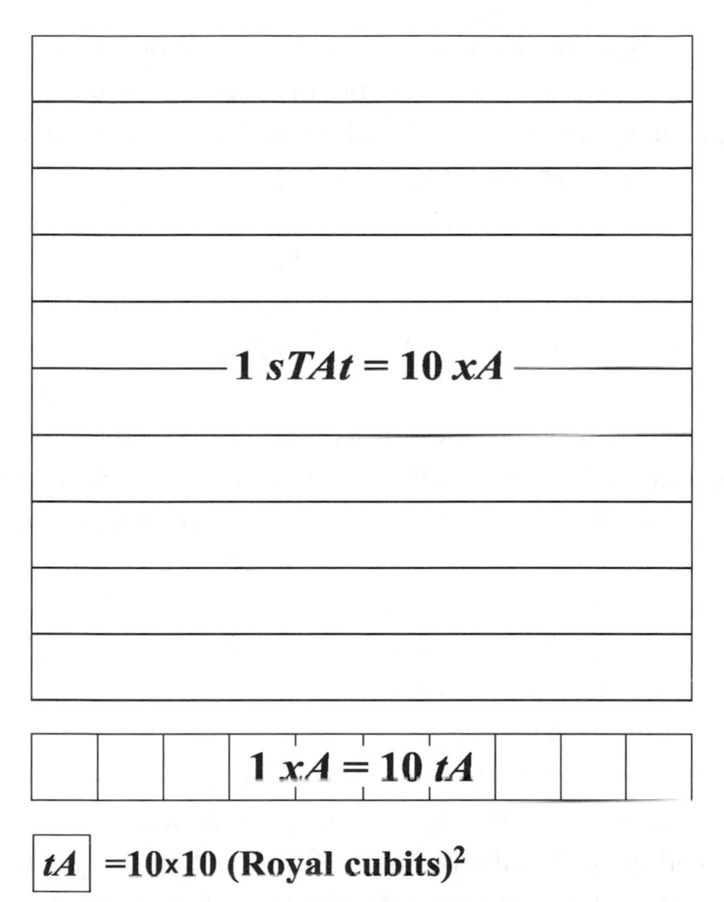

Figure 3.1.1 *Definition of land-area units in ancient Egypt*

Egyptian definitions of volume units also employed base-10:

hin ≅ 0.48 liter
hqAt ≅ 4.8 liters
Xar ≅ 48 liters

This peculiar combination of capital and lowercase letters is just the Egyptologist's way of transcribing hieratic symbols. Vocalization is not much different from how we would attempt to pronounce these letters. Thus, *hqAt* might have sounded something like *hekat*, and it is frequently written as such in English translations.

These base-10 definitions of measuring units are consistent with modern, metric system use, but differ in one important aspect. The metric system requires that all spatial dimensions be expressible in terms of just one unit, the meter:

Length:	10^3 mm =	10^2 cm =		**1 m** =	10^{-3} km
Area:	10^6 mm^2 =	10^4 cm^2 =		**1 m^2** = 10^{-4} hectare =	10^{-6} km^2
Volume:	10^9 mm^3 =	10^6 cm^3 =	10^3 liter =	**1 m^3** =	10^{-9} km^3

This tabulation expresses the essence of the metric system, and the definition of the meter is irrelevant. In practice, replica standards employed by manufacturers of measuring devices assure worldwide uniformity. Today the meter is precisely defined by a wavelength of light, which is in turn determined by fundamental constants of nature that are unchanging, or at least sufficiently unchanging to not affect the everyday life of anybody except physicists.

The original definition of the meter was that the distance from the equator to the poles was 10,000 km. This was a somewhat absurd definition because it was not an easily accessible standard, but it had the political advantage of not relating the meter to the measuring unit of any particular country and made its worldwide acceptance easier. It has the advantage of making the circumference of the earth an easily remembered 40,000 km.

This definition also had the minor consequence of being misunderstood by marginally sane mystics to mean that God created earthly dimensions, rather than that the meter was created by humans to make a dimension of the earth a simple integral multiple of the meter. Thus, a numerically oriented mystic recently discovered that English units are also God-given because the polar diameter of the earth is exactly 500,000,000 inches. In section 4.4 we shall see other examples of this belief that if some measurement is a simple integer multiple of some unit of measurement, it cannot be just coincidental. There is a logical reason for such belief. When we set the dimensions of something we create, we tend to choose simple integer multiples of a unit of measurement. Thus, for example, when creating a window, we would intuitively choose a width of 50 cm rather than 47 cm, unless compelled by some restraint to choose the nonsimple integer. Thus, an earth diameter of 500,000,000 inches is not just coincidental to someone who believes that our universe is a creation of God's intelligent design and that God, like us, also chooses dimensions for his creations that are simple integer multiples of a unit of measurement.

For Egyptian spatial measurements to be a true metric system, the volume units had to be defined only in terms of cubic royal cubits. Rather, as we have just seen, for everyday use Egyptian volumes of most solids and liquids were independently defined in terms of sizes of convenience. However, the Egyptians were aware that volume could be defined in terms of cubic royal cubits, and they did their earth-excavation calculations in such terms. Thus, for engineering calculations they used a completely base-10, metric system. This is very similar to the situation in England before 1965 and in the United States today, where everyday measurements use natural, intuitive English units, while all scientific and most engineering measurements use the metric system.

Operating with a decimal number system and nondecimal measurement units is usually only a minor inconvenience. However, the partial conversion to metric measuring units in the United States has produced a major disaster. In 1999 *Mars Climate Orbiter* crashed into Mars because the engineers who designed the probe used English units while the NASA controllers thought the units were metric. A $125 million mistake! Partial conversion is also no doubt responsible for many nuisance errors, like stripping threads by inserting bolts with metric threads into nuts with English threads, but they just do not make headlines.

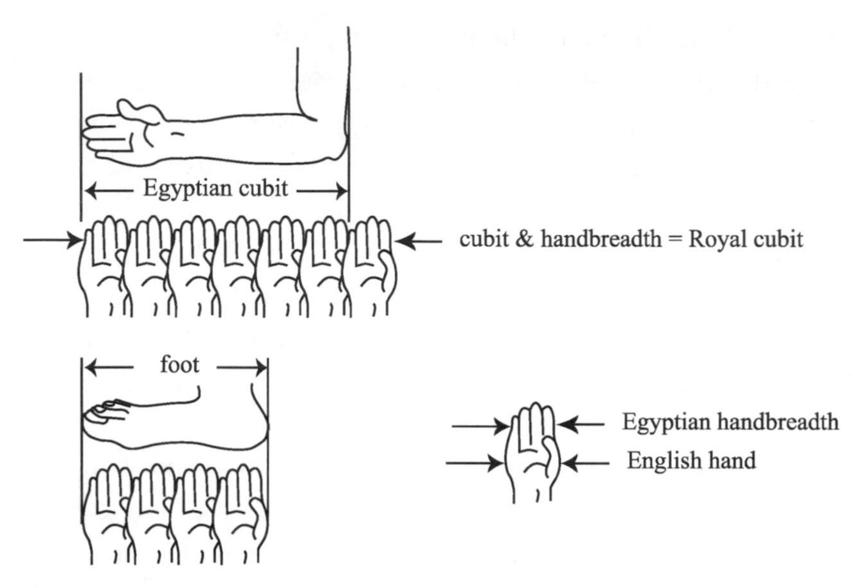

Figure 3.1.2 *Definition of Egyptian royal cubit*

By the time the Great Pyramids were built in the third millennium BCE, the basic Egyptian unit of linear measure was the royal cubit. The word *cubit* is derived from the Latin word for elbow and defines a measurement unit from the elbow to the fingertips, which you can easily test with your own arm. It is a distance of about 18 inches or 457 mm. The cubit is a natural, body-parts unit and is subdivided into natural, body-parts units of 6 handbreadths or 24 fingers. Such subdivision into even numbers with convenient divisors is logical and normal practice in defining measurement units. However, as illustrated in figure 3.1.2, the royal cubit is longer than a cubit by a handbreadth, and is subdivided into 7 handbreadths of 28 fingers.

Figure 3.1.2 also illustrates the natural measurement unit of the Egyptian foot, essentially equal to the English foot, but it is subdivided into 4 hands, each of which is subdivided into 4 fingers, while the English foot is subdivided into different but also natural units of 3 hands, each of which is subdivided into 4 inches (thumb widths). The division of the English foot into 12 inches is a natural consequence of physiology, which use has survived because it fortuitously has many useful divisors. The hand as a unit is long obsolete but is still used to define the height of horses by the tradition-preserving horsey set.

The employment of different replacement schemes for counting and measurement has engendered considerable confusion, not among ancient users, but among historians and archeologists who millennia later have tried to understand the logic of the ancient choices. The lack of appreciation of the existence of these two independent origins of replacement schemes has led to some quite-wrong conclusions. Perhaps the most glaring is the conclusion that the subdivision of the Roman measurement unit of the *pedes* (feet) into 12 *uncia* (inches) reflects base-12 counting. It does not; it reflects human physiology.

English units are related to similar Roman units, which are related to similar Greek units, which are related to similar Egyptian and Babylonian units. However, it is impossible to state that the adoption of similar units was a result of diffusion because, like the adoption of base-10 counting, basing units on the same body parts is also natural and intuitive, and they probably were invented independently in many places.

Not everybody has the same-size body parts, nor do different cultures measure the same body parts in exactly the same manner (compare English and Egyptian definitions of the foot in figure 3.1.2). As different cultures progressed to a state where measurements that were more precise were required, they eventually set measurement standards by government decree. Each culture defined its own standards, and so there were culture-to-culture differences.

The cubit is a natural body-parts unit, and it is naturally divided into 6 hands of 24 fingers with many convenient divisors. So why did the Egyptians define and widely use the royal cubit? It is not the length of any body part, and it is divided into 7 hands or 28 fingers, with fewer convenient divisors. The Greek and Roman successor civilizations did not use the royal cubit, but they did continue to use the cubit with its more convenient divisors. Continued use of the royal cubit into the sixth century BCE is attested to by the biblical quotation:

> Behold . . . there was a man . . . with a line of flax in his hand, and a measuring reed . . . of six cubits by a cubit and a handbreadth.
>
> Ezekiel (Old Testament); Yehezqel (Hebrew Bible), written in sixth century BCE

The interpretation of this somewhat enigmatic passage, although no more enigmatic than most Bible passages, is that the measuring reed was six royal cubits long.

The length of the royal cubit was determined to be 20.62 ± .005 inches (523.7 ± .1 mm) by Flinders Petrie, the father of scientific Egyptology, in a meticulous survey of the Giza Pyramids in 1880. However, such precision is only approached with modern instrumentation that is much more precise than what was available to Petrie, and certainly than what was available to the ancient Egyptians when they constructed the pyramids. Petrie either did some possibly unconscious data selection or did not do his statistics correctly. (Petrie would not be either the first nor the last good-intentioned scientist to do so.) Even granting that a single standard for the royal cubit was used in the construction of all the Giza Pyramids, such precision is not reasonable. Variations among ancient measuring rods that have been found yield a more reasonable definition of 524 ± 2 mm.

But the royal cubit was not used just for building pyramids with perhaps sacred motivation for precision; it was the basic unit used to define land areas for agricultural purposes, and therein lies my explanation for its existence and importance. As we shall see in sections 4.1 and 4.2, multiplication and division by two was the basis of all Egyptian arithmetic. How do you multiply or divide areas by two?

The Egyptian solution was to define two new measurement units: a royal cubit of 7 handbreadths and a remen of 5 handbreadths.

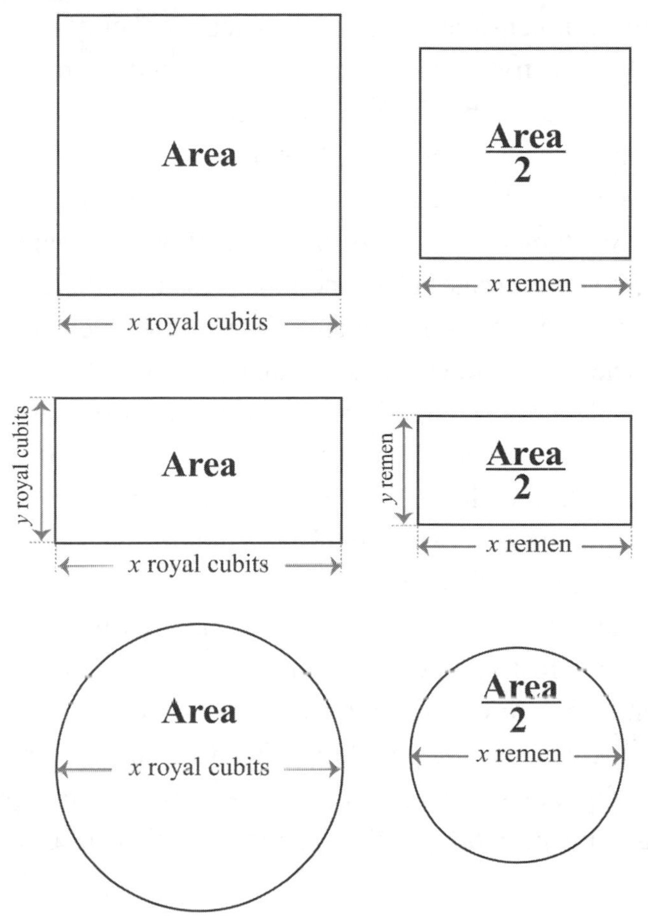

Figure 3.1.3 *Egyptian halving or doubling of areas*

In order to halve an area, Egyptians first measured its linear dimensions as a certain number of royal cubits. A half-size area would be one with its linear dimensions given as the same number of remen. Figure 3.1.3 illustrates this process for various areas (square, rectangle, and circle): by proceeding from left to right, area is reduced by $(5/7)^2 = 25/49 \cong 1/2$.

In order to double an area, they first measured its linear dimensions as a certain number of remen. A doubled area would be one with its linear dimensions given as the same number of royal cubits. Figure 3.1.3 illustrates this process: by proceeding from right to left, area is increased by $(7/5)^2 = 49/25 \cong 2$.

Although not themselves body parts, the royal cubit and the remen are conveniently related as a *cubit and a handbreadth* and as a *cubit less a handbreadth.* Nowadays, with ready availability of inexpensive and accurate rulers and tape measures, there is no need for body-parts rulers as in ancient times. But when required to measure when a tape measure isn't handy, we still step-off a rough measure heal-to-toe in feet. For Americans, who still use English units, the count in feet suffices, but for most of the rest of the world, a nuisance multiplication of feet by 30 cm is required to convert to metric system units. This is one of the few advantages of retaining natural, intuitive English units.

How the Egyptians arrived at the remen/royal-cubic method remains an interesting question. Perhaps it was based on nothing more than a chance observation that $(5/7)^2 \cong 1/2$ and $(7/5)^2 \cong 2$. However, there is also the possibility that it reflects a deeper understanding of geometry than has previously been ascribed to the Egyptians. This possibility will be pursued further in section 4.3.

3.2. HERDER-FARMER AND URBAN CULTURES BY THE WATERS OF BABYLON

In the nineteenth century, the ruins of the Assyrian empire on the upper reaches of the Tigris and Euphrates were the first Mesopotamian sites excavated by archeologists. Thereby, the study of ancient Mesopotamian civilization came to be called Assyriology. Assyria was a brutal military power. But it was in Babylon and its neighboring city-states on the lower reaches of the Tigris and Euphrates, particularly during their glory days from about 2000 to 1600 BCE, that important intellectual contributions were made. With poetic justice, ancient Mesopotamian civilization is now generally referred to as Babylonian. Although a succession of empires dominated ancient Mesopotamia, I shall, as is frequent practice, sometimes just use the word Babylonian and not bother to specify which particular empire happened to dominate at the time.

The Babylonian system of written numbers followed an evolutionary path similar to that of Egypt. However, rather than follow the evolution of the Babylonian number system chronologically, I shall start with the more or less final form of Babylonian numbers as they were written after about 2000 BCE and then work backward to the antecedents.

Egyptian *hieratic* writing evolved where many differently shaped symbols could be conveniently painted on papyrus, a paper-like material from which the word *paper* is derived. Papyrus writing stock was expensive to produce and was used sparingly. Like most organic materials, it tends to disintegrate over time. Egyptian mathematics was recorded in hieratic on papyri, and so very few have survived that it is questionable whether we have a realistic estimate of Egyptian mathematics.

On the other hand, Babylonian *cuneiform* writing evolved where only a limited number of shapes could be made by pressing a wedge-shaped stylus into soft clay tablets, their most accessible writing material. The Babylonians certainly had a less convenient writing stock than the Egyptians did, but for archeologists it was an ideal choice. Clay tablets were easy and cheap to make and were extensively used. After baking in the sun or a kiln, they do not disintegrate with age, and thus the surviving Babylonian mathematical record is extensive, although certainly not complete.

Most historians believe that the mathematics of Babylon far surpassed that of Egypt, but it is questionable whether this evaluation reflects reality or simply reflects the more extensive survival of the Babylonian record.

The Babylonians invented a *positional* number system, a landmark event in mathematics, but with the position values given as powers of 60. Table 3.2.1 shows that cuneiform symbols for the numbers from 1 to 59 are *additive* combinations of just two symbols, a units symbol and a tens symbol. Inscribing a units or a tens symbol on a clay tablet was simply a matter of rotating a wedge-shaped writing stylus by 90 degrees.

1	11	21	31	41	51
2	12	22	32	42	52
3	13	23	33	43	53
4	14	24	34	44	54
5	15	25	35	45	55
6	16	26	36	46	56
7	17	27	37	47	57
8	18	28	38	48	58
9	19	29	39	49	59
10	20	30	40	50	

***Table 3.2.1** Additive cuneiform symbols for numbers 1 to 59*

For easier reading and writing, I shall sometimes transcribe Babylonian cuneiform numbers into *decimally transcribed sexagesimal* notation: Hindu-Arabic symbols with colons separating positions, just as we currently write time as hours:minutes:seconds, our vestigial use of sexagesimal numbers (see section 1.3 for more on transcription of number systems).

There are several conventions in use for transcribing sexagesimal symbols into decimal symbols, and each convention has its pros and cons. In the convention I have chosen, a period denotes a *sexagesimal point*. With this convention, 1.5 could be the decimal expression for 1 1/2 or the sexagesimal expression for 1 1/12 (5/60 = 1/12). When such an ambiguity arises, I shall use subscripts such as 1.5_{10} or 1.5_{60}.

Except for the fact that every position in this base-60 system can have a value up to 59, while every position in the decimal system can have only a value up to 9, cuneiform numbers are theoretically just as easy to read as decimal numbers. But in practice there are difficulties. Using symbols "v" and "<" as easy-to-type representations of the cuneiform units and tens symbols, the cuneiform number (vv vvv <<<<<vv) can be decimally transcribed as 2:3:52, which indicates that it is a three-position number but does not unambiguously define what the positions are. We can write the three positions as n, $n + 1$, and $n + 2$, but n is only known from the context of the document, and so $2{:}3{:}52 = 2 \times 60^{(n+2)} + 3 \times 60^{(n+1)} + 52 \times 60^{n}$. Assuming n to be zero, then $2{:}3{:}52 = (2 \times 60^2) + (3 \times 60) + 52 = 7{,}432$. On the other hand, if the context implied a larger number, then $n = 1$ might be correct and the correct transcription would be 2:3:52:0, and it would have a decimal value of $(2 \times 60^3) + (3 \times 60^2) + (52 \times 60) = (2 \times 216{,}000) + (3 \times 3{,}600) + (52 \times 60) = 445{,}920$. However, if the context implied a fraction, then $n = -3$ might be correct, the transcription would be 0.2:3:52, and it would have a decimal value of $2/60 + 3/60^2 + 52/60^3 = 0.034407$. (In power notation the negative exponent in 60^{-n} means $1/60^n$.)

The Babylonians did not have a sexagesimal point to indicate separation of the integral part of a number from the fractional part; neither did they have a zero symbol. These are added to the decimally transcribed numbers to make the values unambiguous. It is difficult to conceive of using our base-10 system without either a zero symbol or a decimal point, and thus it appears quite amazing that the Babylonians were satisfied for so long with their ambiguous writing system. However, these deficiencies in the Babylonian number system are more in the eyes of modern translators than they were in practice, and instances where $n \neq 0$ were probably rare.

If a Babylonian weighed something as 0.30_{60} talent = 0.5_{10} talent, in the absence of a sexagesimal point it would have been written as <<< and have required knowing from the context that $n = -1$. However, by expressing the weight in terms of a unit that was 1/60th as big, the mina, this weight would have been simply <<< (30) mina with $n = 0$. That units were frequently not inscribed on clay tablets because they were obvious practice to the Babylonians has been very confusing for modern translators.

By about 2000 BCE the Babylonians had a metric system, with base-60 for both their number system and their units of measurement. It was easy to avoid the need for a sexagesimal point just by using a smaller unit. It did not require a change in the digits, just as with modern metric system practice when, for example, writing 175 cm rather than 1.75 m. For users of a decimal number system and English measurement units, which is common practice in the United States today, it is also possible to eliminate the need for a decimal point by choosing a smaller unit, but this requires a nuisance calculation and a change in digits, as, for example, when writing 21 inches rather than 1.75 feet.

We currently use zeros to write large numbers such as 7,000 or 7,000,000, but we can dispense with them by writing 7 thousand or 7 million. This was the Babylonian solution, but this has also been confusing for modern translators since the units were frequently not written because they were obvious to the writer.

There is, however, another difficulty in reading cuneiform numbers due to the lack of a zero symbol. One can interpret a number written as (∨∨ <<<<<∨∨) either as 2:52 or as 2:0:52, because it is not absolutely clear whether the space between adjacent positions is just somewhat larger than usual or whether an empty position is intended. With Egyptian hieroglyphic or hieratic *additive* numbers, a zero symbol had no essential role, and this problem did not arise. Egyptian writing also had distinctive symbols for fractions that obviated the need for a decimal point. Thus, prior to the inventions of a zero symbol and a separation point, additive Egyptian hieroglyphics and hieratics were easier to read and probably led to fewer arithmetic mistakes than positional Babylonian cuneiform. This accounts for why the Egyptian number system never made the transition from additive to positional, a question previously posed in section 1.2.

Notwithstanding these difficulties, such cuneiform numbers satisfied the computational needs of Babylon for some fifteen hundred years. Eventually, around 500 BCE, the Babylonians did invent a symbol to mark an empty position and thereby invented the zero, another landmark event in mathematics. It was not until around the first century that "the best of all possible worlds" evolved in India with the combination of the base-10 and positional concepts to form our presently used *positional*, decimal system. I do not know to what extent this was an original Indian invention or was inspired by ideas from Egypt and Babylon. Note that Egyptian hieratic script already had separate symbols for each of the numbers from 1 to 9 (see table 3.1.2) and the Babylonians already were using positional notation and had an empty-position symbol, a zero.

It is sometimes more convenient to write Babylonian cuneiform numbers as decimally transcribed sexagesimal numbers, but the true character of the Babylonian number system is lost in translation. Abacus addition in number systems with a base originated in a pebble-counting era (see section 1.2). The abacus addition of table 3.2.2 clearly exhibits the true character of Babylonian counting as an alternating sequence of 1-for-10 and 1-for-6 replacements, which produces a sequence of 1-for-60 replacements, a base-60 system.

Although the abacus addition in table 3.2.2 requires only one symbol, the use of two symbols more readily differentiates between columns with 1-for-10 replacements and columns with 1-for-6 replacements, and this is also how the cuneiform numbers in table 3.2.1 are written.

Position value	**60^2**		**60**		**1**	
	x10	**x1**	**x10**	**x1**	**x10**	**x1**
Replacements						
Addend = 58:29 **= 𒐐𒐍 𒎙𒐎**						
Addend = 43:54 **= 𒐏𒐈 𒐐𒐉**						
Sum = 1:42:23 **= 6143_{10} =**		𒁹	𒐏	𒈫	𒎙	𒐈

Table 3.2.2 *Abacus, Babylonian sexagismal system addition*

Adding written Babylonian cuneiform numbers, with erasures of ten ∨ symbols in a column and replacement with one < symbol in the next column, and erasures of six < symbols in one column and replacement with one ∨ symbol in the next column, perfectly mimics abacus addition.

FUN QUESTION 3.2.1: Add the Babylonian cuneiform numbers

𒈫 𒐐𒐉 𒐏𒐈 and 𒁹 𒌋𒐊 𒐐𒐋

in both cuneiform and decimally transcribed symbols.

Babylonian cuneiform can be described as *abacus notation.* Except for two simple rules for replacement, addition with Babylonian cuneiform requires no memorization of addition tables. True sexagesimal addition with 59 symbols would have required addition tables with 1,770 entries (see table 1.3.4). Archeologists have not found any such tables, and so abacus addition was certainly used.

Usually, the larger the base, the more compact the notation (see section 1.3). However, this is not the case for Babylonian cuneiform. This is demonstrated by comparing abacus counting for base-10 and Babylonian base-60. By comparing abacus exhibitions, we eliminate differences in notation. Table 1.2.1 presented a decimal-abacus exhibition of the four-digit number 5,829 in four columns, which we can write as $(58 \times 100) + 29$; table 3.2.2 presents a Babylonian sexagesimal-abacus exhibition of the same four digits in four columns, but now the decimal value of 58:29 is only $(58 \times 60) + 29 = 3{,}509$.

Did Babylonian cuneiform originate as base-60 with subsequent division of each position into two columns, as in table 3.2.2, to ingeniously enable easy abacus addition, or did it originate as a sequence of alternating 1-for-10 and 1-for-6 replacements with its resulting near-equivalence to base-60 just an unintended accident? If we refer back to the invention of pebble counting with replacement (see section 1.2), we see that the purpose of replacement was to better visualize the number of pebbles in a count. Only small replacement numbers have this property, and thus we can be sure that base-60 counting was never chosen; it just happened.

Now let us return to a question previously posed (see FUN QUESTION 1.3.7): Why did the Babylonians choose a base-60 number system? We now know that this is the wrong question. The right question is why they chose a sequence of alternating 1-for-10 and 1-for-6 replacements and not a sequence of just 1-for-10 replacements. Misguided by the use of decimally transcribed cuneiform that completely obscures the alternating replacement sequence, historians have always asked the wrong question, but it is amusing to look at some of their conjectures.

Already in the fourth century, Theon of Alexandria hypothesized that 60 was chosen as a base because it has many divisors: 1, 2, 3, 4, 5, 6, 10, 12, 15, 20, 30, 60. Since division is a difficult arithmetic operation (or at least it was until the invention of the electronic calculator), it is convenient to use a base with many divisors so that some common divisions are easy.

To illustrate the advantage of many divisors, note present-day subdivision of a dollar into one hundred cents, essentially base-100 use that provides many useful divisors. However, the primary reason for decimalizing currency is not to provide many useful divisors, but is to make calculation easier with the long-adopted decimal number system. The first nation to realize the advantage of a decimalized currency and to decimalize its currency was the United States in 1792. Ironically, the United States is the only major nation that has not adopted the metric system of decimalized measuring units, even though Thomas Jefferson had proposed decimalized standards of weights and measures in 1791.

FUN QUESTION 3.2.2: Prior to converting to decimal currency in 1971, the system in England was 12 pence = 1 shilling, 20 shillings = 1 pound, and therefore 240 pence = 1 pound. This is also a system with many convenient divisors. What are the divisors of 240?

A somewhat mystical proposal was based on Babylonian division of a year into 360 days, with five noncounted holy days added at the end of the year. Such usage is the origin of the modern division of a circle into 360 degrees. Why did they not simply define a year as 365 days? Because 360 is 6×60; and six equilateral triangles, six *perfect* triangles (that is the mystical part), form a hexagon, which can be inscribed in a 360-day calendar-circle, thereby making 60 a natural unit for counting.

A purely mystical proposal for the origin of base-60 was that Babylonian numerology assigned numbers to each of their gods. To their celestial god they assigned the number 60. He was the father of the earth god whose number was 50, and so on. They most certainly assigned such numbers to gods after base-60 had been in use.

Among other hypotheses are some that proposed base-60 was the result of a union between cultures. One proposal was that one of the cultures used base-10 and the other base-6. Another proposal was that one of the cultures used base-12 and the other base-5. Base-60 was presumably chosen because it contained all of the divisors of the number bases of both cultures. Such a diplomatic solution would hardly be expected today, much less in less civilized times. Neither is there any evidence that any culture ever used base-12 (see section 3.1) or base-6.

The solution to the base-60 enigma lies in a pebble-counting era, long before the invention of writing and long before sophisticated thinking about bases or divisors. So let us go back to long before 2000 BCE and look at the archeological evidence.

In Mesopotamia, *clay counters* of different sizes and shapes (spheres, disks, cones, cylinders, and so on) replaced pebbles. Thousands of these *counters* have been found in archeological digs throughout the Near East and dated from the ninth millennium BCE to the second millennium BCE. This essentially brackets the period from the beginning of the Neolithic Revolution to the emergence of a class of scribes who did written arithmetic. At archeological sites in Sumer, the predecessor state to Babylon, *counters* were found dating from around 3500 BCE that could be shown not to follow the natural and intuitive base-10 replacement sequence. Rather, they assumed the *replacement* sequence illustrated in figure 3.2.1.

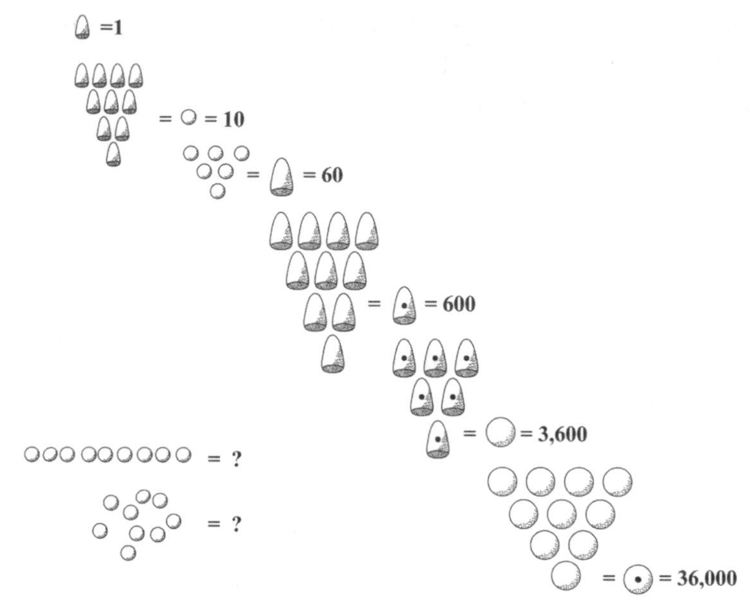

Figure 3.2.1 *Counter use in Sumer circa 3500 BCE*

Someone in Sumer, using *counters* to count something, rather than using just 1-for-10 replacements, preferred to use an alternating sequence of 1-for-10 and 1-for-6 replacements. Unaware though he surely was, he had invented an *additive* base-60 counting system.

Figure 3.2.1 also illustrates what has been called the "rule of four." With just a glance, the eye cannot discern the number of objects in a group if the number is greater than four. In order to discern the number of *counters* quickly, they are arranged in an ordered pattern where each grouping never contains more than four *counters.* In the lower left corner of figure 3.2.1 are two groupings of more than four *counters*; one is an ordered line and the other a more or less random array. You can readily test that you cannot discern the number with just a quick glance at either of these groupings. The cuneiform numbers of table 3.2.1 follow the rule of four. So does the way we nowadays write telephone numbers, credit card numbers, and other numbers we need to read quickly and accurately.

We must first answer how we know that this really was the replacement procedure. Differently shaped clay *counters* do not define the replacement procedure; their existence only shows that there was some undefined replacement procedure. Fortunately, with the invention of writing around 3000 BCE, written symbols mimicked the shapes of *counters*. The values of the written symbols could be determined from the context of the document, and hence values could be assigned to the mimicked *counters*. Table 3.2.3 shows the evolution from *counters* to mimicking symbols and then to cuneiform in Sumer. The archeological evidence from Sumer definitively proves that the number-system base was determined during the prewriting, pebble-counting era.

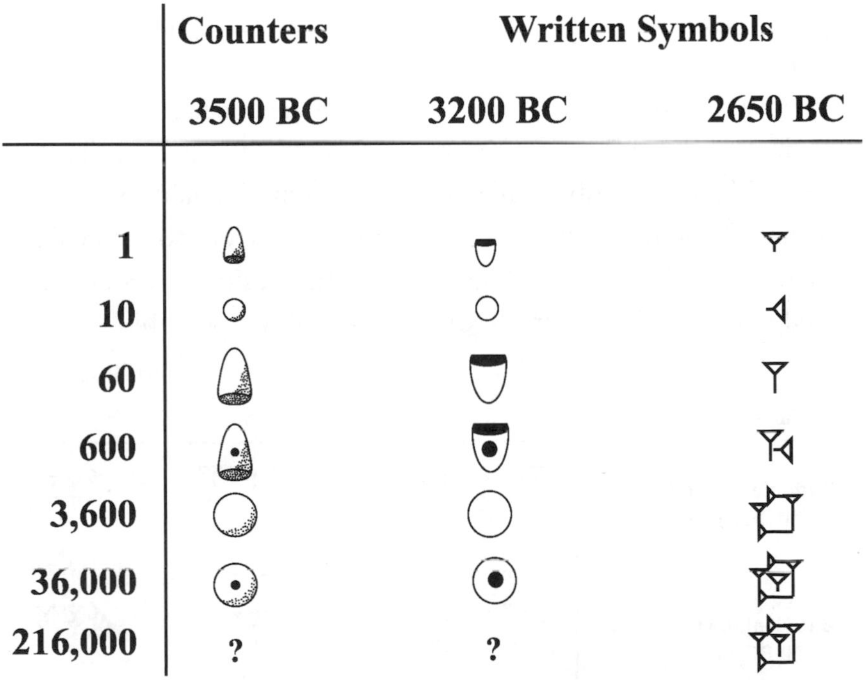

	Counters	**Written Symbols**	
	3500 BC	**3200 BC**	**2650 BC**
1			
10			
60			
600			
3,600			
36,000			
216,000	?	?	

Table 3.2.3 *The evolution of number symbols in Sumer*

Note how the symbol for 600 combines the symbols for 10 and 60 and the symbol for 36,000 combines the symbols for 10 and 3,600. This is a clever way of not having a proliferation of shapes/sizes, which would be confusing. It also distinguishes the 1-for-10 replacement counters from the 1-for-6 replacement counters. Such subtle touches in the design of the counters imply that the 3500 BCE design represents evolution over a long period.

In section 3.1, I conjectured that base-10 of the Egyptian hieroglyphic number system was determined in the prewriting, pebble-counting era. We had reasonable evidence—the counting methods of present-day, primitive herder-farmers—but there was no archeological evidence from the ancient era. We now also have that.

Table 3.2.3 also shows that even after Sumerian numbers had evolved into cuneiform symbols, they were still *additive*. In principle, it is possible to write additive number symbols in any sequence, but in practice, they were usually written in a logical sequence: horizontally, with symbol-values increasing from right to left, just as we currently write base-10 numbers. Around the year 1900 BCE, some scribe realized that the Sumerian additive system was unnecessarily redundant, with both the symbol and the position able to define the value, and he invented the *positional* system illustrated in table 3.2.4. Prior use of abacus addition probably played a role in inspiring the adoption of this two-symbol, positional system.

Decimal transcription	**1 : 2 : 34 : 56**
Sumerian cuneiform (2650 BCE)	
Babylonian cuneiform (1900 BCE)	

Table 3.2.4 *The transition of numbers from Sumerian additive to Babylonian positional*

Around the year 2000 BCE, Amorites migrating from the Arabian Desert conquered Sumer and founded Babylon. The Babylonians adopted much of the rich Sumerian culture, including cuneiform writing. The Amorites were of the Semitic language group, variations of which were spoken in Arabia, Canaan, and Egypt (and still are today), so understanding each other's spoken language was relatively easy. For international written communication, cuneiform writing was adopted throughout the Near East. This explains why a rich trove of cuneiform clay tablets was found in Tel Amarna, Egypt, the remains of a new capital founded by the pharaoh Akhenaton in Dynasty XVIII (1372–1355 BCE). After about 2000 BCE, there was considerable diffusion of ideas among the cultures of the Near East, and we shall later see evidence for the diffusion of mathematics between Egypt, Babylon, and India. However, during most of the pebble-counting era, neither horse nor camel had been domesticated yet and distant interaction was considerably less. Thus, the completely different number systems in Egypt and Babylon evolved independently, with each culture benefiting little from the other's progress.

FUN QUESTION 3.2.3: What is the decimal value of the number in table 3.2.4?

ANSWER: $60^{3+n} + (2 \times 60^{2+n}) + (34 \times 60^{1+n}) + (56 \times 60^n) =$ $(216{,}000 + 7{,}200 + 2{,}040 + 56)60^n = 225{,}296 \times 60^n$, where n is unknown and can only be known from the context in which the number appears.

The answer to this FUN QUESTION uses a very useful, but hitherto undefined, property of power notation: $60^{3+n} = 60^3 \times 60^n$ and $60^{2+n} = 60^2 \times 60^n$, so 60^n can be factored out of the answer. This can be easily understood by noting that $b^3 = bbb$, $b^2 = bb$, and $b^5 = bbbbb$, so $b^5 = b^3 \times b^2$, and more generally, $b^{m+n} = b^m \times b^n$, QED.

FUN QUESTION 3.2.4: Prove that $b^m/b^n = b^{m-n}$.

FUN QUESTION 3.2.5: Prove that $\sqrt{b^m} = b^{m/2}$.

Now let us return to our newly defined enigma of why six was inserted in the *counters replacement* sequence. Pebble counting in Babylon started, just as it did in Egypt, with natural and intuitive 1-for-10 replacements. Counting continued normally up to a count of 59, and then 1-for-6 replacements alternated with 1-for-10 replacements. We can be sure that the reason for not continuing with just 1-for-10 replacements had nothing to do with improving counting efficiency because nothing could be more natural, more intuitive, simpler, or more efficient than only 1-for-10 replacements. We are sure of this because, of all the various bases that have been invented, base-10 is the only survivor for common use because it has been the fittest.

The innocuous insertion of 1-for-6 replacements in the replacement sequence had the unintended consequence of changing the base from 10 to 60. There are only two possible explanations for this insertion of 1-for-6 replacements: either six was a number that was ennobled in some mystical way and the Babylonians wanted their number system "blessed" by its inclusion, or there was some practical advantage in its use.

* * * * *

Let us consider a mystical explanation first. Mystical explanations are not new, and I have noted some theories earlier in this chapter. It is difficult for us today to appreciate the intrusion of mysticism into the construction of a number system, but for the number system of the inventors of numerology, we must consider this possibility and look for evidence that six was for them a sufficiently ennobled number.

Even in today's more or less rational and scientific world, the concept of a lucky number still exists. I cannot recall any number that is nowadays commonly considered lucky, but thirteen is commonly considered unlucky, although few can explain why it has such a reputation. It is such a prevalent superstition that many modern, high-rise buildings delete thirteen in the floor-numbering sequence. I have never understood how people who live or work on the fourteenth floor in such buildings do not realize that they are really on the thirteenth floor and are doomed. Just as we have deleted thirteen from the floor-numbering sequence, it is possible the Babylonians considered six to be such a lucky number that they inserted it into their counting sequence.

We start our search for an ennobled six in an unlikely spot, the middle of the flag of the present-day State of Israel. Emblazoned there is the six-pointed Star of David. Legend has it that this was the shape of King David's magical shield. In Hebrew, this symbol is called *Magen David*, literally the Shield of David. Although Israel has possibly been more intensively excavated archeologically than any other country, not a single artifact has ever been found decorated with this important Jewish symbol that could be dated to before the sixth century BCE. The explanation is simply that the legend about King David's shield is pure fantasy, invented much later than his reign of around 1000 BCE.

In 604 BCE, King Nebuchadnezzar sacked Jerusalem and deported many Jews to Babylon. After an exile of some seventy years, the Jews were allowed to return, and they brought back with them many stories now included in the Hebrew Bible (essentially the Christian Old Testament) and the Babylonian symbol, the six-pointed star. Many artifacts have been found in Babylon with this symbol, predating the Jewish exile there by thousands of years.

Cultural anthropologists have long understood the ancient symbolism represented by the six-pointed star. The triangle pointing down represents the female; and of course, the triangle pointing up represents the male. One interpretation of the joining of these two symbols to form the six-pointed star is that the star symbolizes unity and harmony. Similar mystical symbolism persists even today in a Hindu meditation device of ancient origin, the *sriyantra*, illustrated in figure 3.2.2. The existence of such symbolism in Christianity plays an important role in the controversial and popular novel *The Da Vinci Code*, by Dan Brown.

Figure 3.2.2 *Sriyantra, Hindu meditation device*

The Babylonians certainly did include numerological considerations in their daily lives. They invented the seven-day week, but because seven was for them an unlucky number, they made the seventh day a day of rest. They now had a six-day workweek, so again six appears in a numerological context, but here it seems to be accidental without any implication of ennoblement.

Incidentally, the Jews in Canaan adopted the Babylonian seven-day week, and some imaginative author gave a uniquely Jewish theological spin to the custom of a weekly day of rest. He made it part of the biblical creation story and elevated its observance to one of the Ten Commandments. The Jewish Sabbath is Saturday. To express their uniqueness, Christians made Sunday their Sabbath, and so, thanks to Babylonian numerology and a little theological politics, we now enjoy a two-day weekend.

FUN QUESTION 3.2.6: If the pebble replacement scheme is alternating 1-for-10 and 1-for-3, what base does it define?

Although numerology could account for the insertion of six in the Babylonian *counters replacement* sequence, nothing appears to make six so uniquely ennobled that it justified disrupting the natural, intuitive use of just a sequence of 1-for-10 replacements. So now, let us seek a practical reason for inserting six in the counters replacement sequence.

* * * * *

We have already seen in section 3.1 that while a sequence of 1-for-10 replacements is natural for counting, other replacement sequences are natural for measurements. It is in the interaction between demands for convenient counting and demands for convenient measuring that we shall seek the answer to the base-60 enigma.

We are all now somewhat aware of this interaction because many countries are currently undergoing a process, particularly slow and painful in the United States, of conversion to the metric system for measurement units. France invented and adopted the metric system in 1793 during the French Revolution. However, clamor by people for their old, familiar units resulted in Napoleon rescinding the adoption in 1812. France readopted it for good in 1837. The conversion to the metric system from familiar, natural, intuitive units of measurement has been a difficult process wherever it has occurred.

Mathematician Oystein Ore in 1948 was possibly the first to conjecture that the base-60 number system was derived from measurement units. He suggested from weight units:

1 *gin* (shekel) ≅ 8 1/3 gm
60 *gin* (shekel) = 1 *mana* (mina) ≅ 500 gm
60 *mana* = 1 *gu* (talent) ≅ 30 kg

The units in parentheses are how these units are popularly known; it is how these units are referred to in the Old Testament.

These weight units are clearly a case of definition after the base-60 number system was in place. The replacements are just 1-for-60, but for this to be the origin of base-60 counting, we require a sequence of alternating 1-for-10 and 1-for-6 replacements, so we can rule out weight units as the origin of base-60. Actually, ascribing the base-60 number system to a base-60 system of weights solves nothing; it simply transfers the proof of why a base-60 system was chosen from numbers to weights.

Weight and value of metals are related, and so money units have often taken the same names as weight units. Thus, for example, money in England is denominated in pounds and in Israel in shekels. Fortunately, when modern Israel chose to call its currency by the biblical name, nostalgia did not override common sense, and it was subdivided into one hundred units in the modern manner rather than into sixty units in the biblical manner.

We turn our search for the origin of base-60 counting to spatial measurements. In Babylon, the same body parts defined spatial units as in Egypt; however, the specifications were somewhat different. Perhaps the oldest preserved standard of length was one inscribed around 2575 BCE on a statue of King Gudea of Lagash, which defines the Sumerian cubit as 495 mm. This value is between the value for the Egyptian cubit of 449 mm and the value for the

Egyptian royal cubit of 524 mm. It is not known whether the Babylonian cubit was derived from the Egyptian cubit or royal cubit, or was an independent definition. Taking into account the uncertainty in the definition of the Babylonian cubit, its value is generally approximated as 500 mm. Another difference between Egyptian and Babylonian cubits is that the latter is subdivided into thirty digits.

Body-parts definitions of measurement units are not of interest here because physiology defines them. What is of interest is how longer units were defined. For construction purposes, every culture defined new units that are small extensions of body-parts units. As might be expected, there is more variation in such extended units because they involve both disparate measures of body-part sizes and a decision about the number of their multiples, as illustrated in the tabulation for some different cultures in table 3.2.5.

English units	1 rod = 15 foot ≅ 4.5 m (pre-sixteenth century definition)
Roman units	1 *passus* (double pace) = 5 *pes* (foot) ≅ 1.5 m
Babylonian units	1 *gi* (reed) = 6 *kush* (cubit) ≅ 3 m, 1 *nindan* (rod) = 2 *gi* ≅ 6 m

Table 3.2.5 *Definitions of extended body-parts units*

The rod, which appears in both English and Babylonian units, is rather too long to carry conveniently, and so it might not have literally been a rod but might have been a rope. Metric units are not included in this tabulation because they are not natural, intuitive units and only serve as a common basis for comparison.

All cultures also had land-measurement units for longer distances that originally had no relationship to body parts, but were frequently defined by farming practice. The most obvious is the English furlong, literally a furrow long. Such units were not precisely defined, but depended on local tradition and farming practice. Eventually, as a culture matures and requires more precision, there is then reconciliation between the body-parts definitions and the farming-practice definitions.

Table 3.2.6 puts such reconciliations into historical perspective by exhibiting the replacement numbers for English, Roman, Greek, and Babylonian distance-measurement units. In this table, the first column at the left is the unit of a foot in the various languages. All these cultures had a foot unit of more or less the same size, and so this is a convenient way of comparing units. Any unit can be expressed in terms of any other unit as the product of all replacements between the two units.

Let us start with the second entry in table 3.2.6, English practice in the fifteenth century. The body-parts unit of the foot had been reconciled with the farming-practice unit of the furlong, but the retained Roman definition of the mile as 5,000 feet had the undesirable property of not being an integral multiple of the furlong. Returning to the first entry in the table, the sixteenth-century reconciliation of units, English surveyors introduced a new unit, the *chain*. The furlong was now defined as a 1-for-10 replacement of chains. The chain was further defined as a 1-for-10 replacement of chain-links. After the sixteenth century, surveyors could now do most calculations in English units with simple base-10 arithmetic using the units chain-link (6.6 feet), chain (66 feet), and furlong (660 feet).

In the evolution of English units, we can observe both the reconciliation between body-parts units and farming-practice units before the sixteenth century and at least a partial reconciliation between the replacements of decimal counting and the replacements of distance-measurement units in the sixteenth century.

The English furlong was clearly derived from the Roman *sta-*

English units after the sixteenth century and reconciliation of furlong and mile												
foot			6.6	link		10	chain		10	furlong	8	mile
furlong = 6.6x10x10 = 660 foot, mile = 8x660 = 5,280 foot												

English units before the sixteenth century, but after reconciliation of foot and furlong												
foot					15	rod			40	furlong	8.3...	mile
furlong = 15x40 = 600 foot, mile = 5,000 foot												

Roman units after reconciliation of passus and stadium												
pes		5	passus						125	stadium	8	mille
stadium = 5x125 = 625 pes, mille = 125x8 =1,000 passus = 5x1,000 = 5,000 pes												

Greek units after reconciliation of pous and stadion												
pous				10	akaina		10	pletheron	6	stadion		
stadion = 10x10x6 = 600 pous												

Babylonian units after reconciliation of nindan and USH												
(foot)	5/3	kush			12	nindan		10	eshe	6	USH	
nindan = (5/3)x12 = 20 (foot), USH = 10x6 = 60 nindan = 1,200 (foot)												

***Table 3.2.6** Replacements for distance-measurement units*

dium, which was in turn derived from the similarly defined Greek *stadion*. The Greek and Roman words obviously relate to the length of a sports arena. Although the agricultural origins of the units were lost in the Roman and Greek names, they were agricultural units. A modern adoption of an agricultural unit for a sports-arena dimension is the cricket pitch that is 66 feet or a tenth of a furlong. Moreover, in the United States we can currently observe the genesis of a new sports-inspired area unit, the *football field*. The agriculture-based English unit of area, the *acre*, is meaningless to modern urban dwellers in the United States. Thus, a reporter describing an aircraft carrier invariably writes that its deck size is three *football fields*, which is easily visualized, even though referring to the size as three *acres* would be just as good an approximation (*football field* = 5,000 yd^2, *acre* = 4,800 yd^2).

The Roman units reflect the importance of administering their far- flung empire. By choosing the *passus* as a unit, the standard military stride became a reasonably precise unit for determining the convenient base-10 unit, the *mille*. To accommodate this definition, the Romans reconciled it with the *stadium* by changing it to 125 *passus* so that the *mille* became a simple 1-for-8 replacement of the *stadium*.

In order to be able to calculate how long a march would take, another unit was defined as the distance marched in one hour. This evolved into the English *league*, which was equal to about three *miles*. We are still familiar with this obsolete unit because of its appearance in the title of Jules Verne's classic *Twenty Thousand Leagues under the Sea*. If we take Verne's use of the word literally, 20,000 leagues are about 60,000 miles. Since the diameter of the earth is only about 8,000 miles, Verne used quite a bit of poetic license to use a word that conveyed more mystery than a more reasonable 20,000 feet.

The Greek units also exhibit reconciliation of the body-parts unit of the *pous* with the agricultural unit of the *stadion*. Measurement up to a *pletheron* is pure base-10, and then comes a 1-for-6 replacement to reconcile the *pletheron* with the agricultural unit, the *stadion*. The Greek measurements started out as a sequence of natural, intuitive 1-for-10 replacements, but had to be followed by a 1-for-6 replacement to reconcile a natural, intuitive, body-parts unit with a natural, intuitive, agricultural unit.

When we finally arrive at the predecessor, in Babylonian units we see a 1-for-10 replacement followed by a 1-for-6 replacement to reconcile the extended, body-parts unit, the *nindan*, with the agricultural unit, the *USH*. The *USH* is twice as long as the *stadion* (≅ furlong). Apparently, the half-*USH* was the original ancestor of the furlong. A possible explanation for this peculiar *stadion* ≅ half-*USH* relationship is that a frequently used Babylonian unit of land area was the *bur*, an area of 1/2 *USH* by 1 *USH* (see figure 3.2.4).

The furlong has remained a standard of approximately 600 feet over an incredibly long six millennia, from the beginning of plowing with oxen around 4000 BCE up to the twentieth century when the tractor radically changed farming practice.

I conjecture that, to reconcile their number and distance-measuring systems, rather than modifying their measuring units, as the English surveyors did in the sixteenth century and as the French did by adopting the metric system in the nineteenth century, the Babylonians modified their number system and retained their measuring units. The counting system in the pebble-counting, prewriting era was not entrenched in extensive archives, so it was probably easier to modify the counting system than the measuring system. Modern experience with the metric system makes us aware of how painful it is to change measuring units. The natural tendency to make things (fields, houses, baskets, jars, tools, etc.) as simple multiples of measurement units means that units are literally built into a culture. A change in measuring units means that there will inevitably be a problem with fitting new pegs into old holes.

At the time, the reconciliation between counting and measurement systems that unintentionally led to the Babylonian base-60 number system was a logical choice. Only centuries later would the awkwardness of arithmetic owing to such a large base become apparent. By then base-60 was so entrenched in Babylonian culture that they were stuck with it.

Using the counter shapes/sizes previously presented in figure 3.2.1, table 3.2.7 shows how the counter displays of various distances depend on how the counters are defined. This table shows

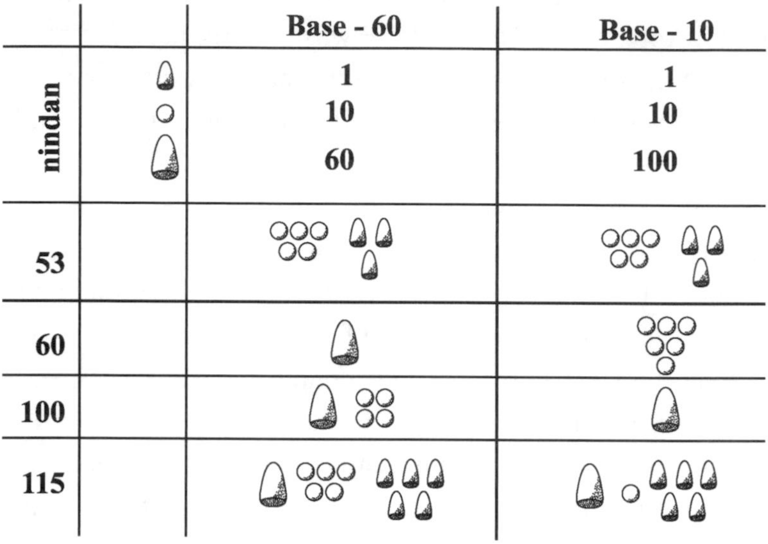

		Base - 60	Base - 10
nindan		1 10 60	1 10 100
53			
60			
100			
115			

Table 3.2.7 *Distance measures for alternative base-60 and base-10 definitions of counters*

that there is no obvious advantage in terms of a simplified display by matching the Babylonian counting system to the measuring system. For any distance up to 59 *nindan*, both base-60 (alternating 1-for-10 and 1-for-6 replacements) and base-10 (only 1-for-10 replacements) produce exactly the same display. For distances that are multiples of 60 *nindan*, base-60 obviously produces a simpler display; for distances that are multiples of 10 *nindan*, base-10 obviously produces the simpler display.

The base-60 advantage is not obvious, but it is real and it is important. Taking the display of counters for 115 *nindan* as an example, the base-60 display can be read directly as mixed units, the everyday way of expressing measurements: 1 *USH*, 5 *eshe*, and 5 *nindan*. The shape/size of the counters defined only an abstract number. The context defined the use and hence the units.

Using counters defined by base-10 requires a nuisance calculation to convert to mixed units. However, the major deficiency of a base-10 display can be seen by again using the 115-*nindan* length as an example. To express it in the adopted mixed units would have required the display:

But before 3000 BCE, writing had not been invented yet, so there would have been no way of knowing how to assign mixed units to this set of eleven identical counters.

The modern metric system was invented to make measuring units compatible with long-adopted base-10 counting. In the prewriting, ancient world, metric-like systems with the same replacement sequence for counting systems and measuring units were also required for efficient recording of measurements. Egypt did it by defining measuring units with the same 1-for-10 replacement sequence used for counting. Babylon did it by defining counting with the same alternating 1-for-10 and 1-for-6 replacement sequence used for distance-measuring units.

FUN QUESTION 3.2.7: Convert a length of 874 *nindan* into Babylonian mixed units (*USH*, *eshe*, *nindan*).

We now understand why the Babylonians included a 1-for-10 replacement followed by a 1-for-6 replacement in their number system. But why did they continue their number system with a sequence of 1-for-10 and 1-for-6 replacements? One possibility is that this was just an instinctive thing to do and not much thought was involved. However, the motivation for extending the alternating sequence of replacements was probably their land-area measurement units.

We previously noted how ancient Egyptians perceived land-area units as generated one dimension at a time (see figure 3.1.2).

English land-area units (and the predecessor Roman and Greek units) were similarly perceived as generated one dimension at a time. Figure 3.2.3 illustrates how, starting with a square rod, a rood (long obsolete), then an acre, and finally a square furlong are generated. This appears to be a common, intuitive perception, and we can reasonably expect Babylonian definitions of area units were generated in the same manner.

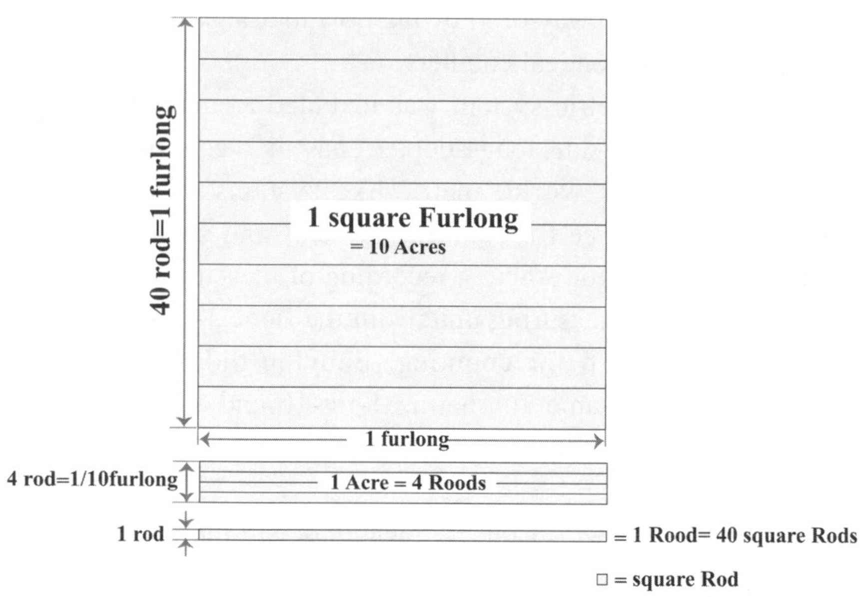

Figure 3.2.3 *Definition of land-area units in English units*

Figure 3.2.4 presents definitions of Babylonian land-area units. There is some confusing nomenclature because the units of distance, the *eshe* and the *USH*, also appear as area units because the concept of generating a line did not start with a mathematical point but with a finite square, defining both a distance and an area. Presumably, when used in context, the Babylonians were able to distinguish between distance and area measures. Perhaps the word *eshe* had some meaning that implied replacement by ten and the word *USH* had some meaning that implied replacement by six. Figures 3.2.3 and

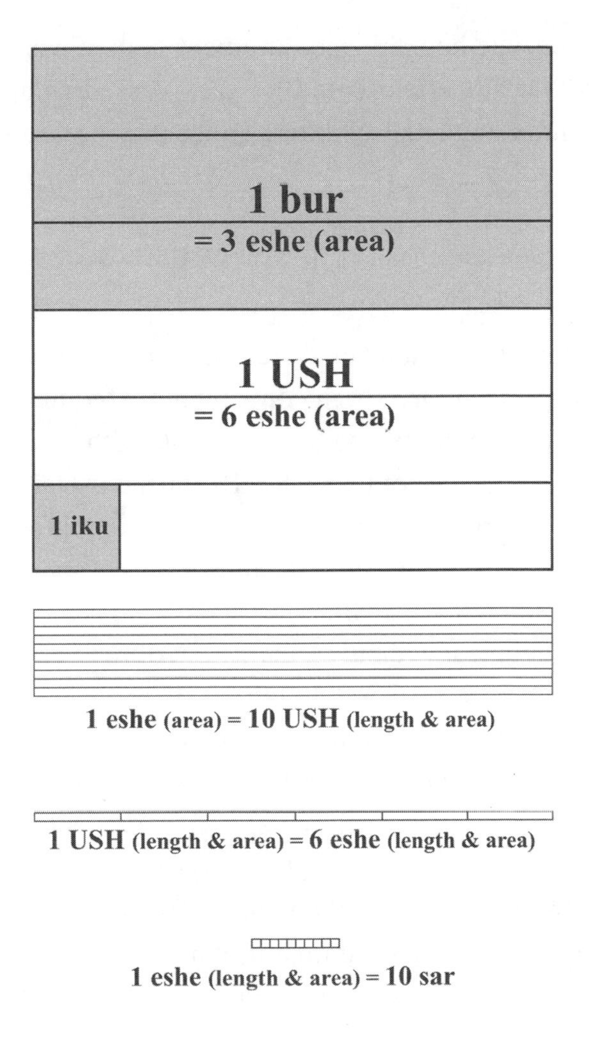

Figure 3.2.4 *Definition of land-area units in Babylon*

3.2.4 are amazingly alike and only differ in that the generation of the rood from 40 square rods is done in one step while the generation of the *USH* (length & area) from 60 *sar* is by the two-step, 1-for-10 and 1-for-6 replacement sequence.

To appreciate how this measurement system worked, let's consider how the Sumerians could have expressed an area of $4{,}321_{10}$ square *nindan* with the counters presented previously in figure 3.2.1:

4,321 square *nindan* = [counters] = 1USH (area), 1 *eshe* (area), 2 *USH* (length and area), 1 *sar*.

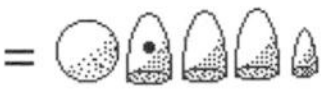

Again, a compatible number system and measurement units allow counters to be read directly in convenient mixed units.

For the number system to be compatible with land-area measurement units, a sequence of two 1-for-10 and 1-for-6 alternations were required. Similarly, for the number system to conform to volume-measurement units, a sequence of three 1-for-10 and 1-for-6 alternations were required. Thereafter, the alternating sequence was apparently just continued, without visualization of real-world dimensions.

Sometime between 3500 and 2000 BCE, it was realized that it was unnecessary to think in terms of alternating 1-for-10 and 1-for-6 replacements, but that a base-60 number system encompassed and extended the originating replacements concept. Only vestiges of the originating replacements concept survived in the way numbers from 1 to 59 were written and in the retaining of some former units such as the *eshe* and *USH*. By the time use of metals became widespread and weight units were required, units were defined strictly as base-60, as we have seen.

Area measurements are the most pertinent for agricultural purposes, and table 3.2.8 gives the legendary origins for the basic units for different cultures. One conclusion that can be derived from this tabulation is that either English oxen were about 50 percent more vigorous than Roman oxen or legends are not to be taken literally. Favoring the latter interpretation, the tabulation shows that an area of about the size of an English acre was widely used, both geographically and temporally, to define a parcel of agricultural land.

Furthermore, land area was originally defined by how much a man could plow in a day, or by some other agricultural criterion, which eventually had to be reconciled with body-parts units. Since the use of oxen for plowing dates from the fourth millennium BCE, we can date the origin of parceling land into approximately acre-size units from the pebble-counting era.

English units	acre	area a yoke of two oxen can plow in one day = 4,047 m^2
Roman units	*jugerum*	area a yoke of two oxen can plow in one day = 2,530 m^2
Babylonian units	*iku* (*eshe*/6)	area enclosed by irrigation dike = 3,600 m^2

***Table 3.2.8** Definitions of agricultural-area units*

Prior to around 3500 BCE and the defining of the number system by counters with an alternating 1-for-10 and 1-for-6 replacement sequence, many other counter shapes and replacement number sequences had been used. Particularly relevant is that animals were originally counted with the natural, intuitive sequence of 1-for-10 replacements; counting of *things* had been base-10. Thus, for example, the counter shapes and values for counting horses were

▯=1, ◯=10, ⧗=100.

The combination of symbols

⧗◯◯▯▯▯

represented 123 horses.

When the replacement sequence represented measurement units, there were different counter shapes and replacement numbers for different measurements. Thus, for example, the counters for measuring volumes of barley seed were

◯= 1, ⛉ =5, ○ = 30.

This sequence of volume units has no obvious mathematical logic, and indeed it should not have. It simply represents convenient volumes governed by existing technology. Perhaps the one-unit and five-unit volumes represent pottery vessels and the thirty-unit volume represents a clay-lined pit.

The combination of counters represents $30 + (2 \times 5) + 3 = 43$ volume units of barley seed.

Counter shapes and replacement numbers for grain (seed) volumes were also used to define land areas, both a logical and a widely used convention in the ancient world. Thus, this counter display could also possibly have defined 43 land units as 43 *iku* and one volume unit of barley seed would have sown an area of one *iku,* the standard size of an irrigated plot.

FUN QUESTION 3.2.8: Using the base-60 counters of table 3.2.2, how would 123 horses be expressed? How would 43 *iku* be expressed? What is the replacement number sequence?

= 1, = 5, = 30.

Most of the excavated counters were found in ruins of temples and palaces, and were apparently receipts for taxes paid. The counters were not currency, just accounting records, and had no intrinsic value. Many counters have also been found deposited in graves. It was ancient practice to deposit a man's tools-of-his-trade in his grave for his use in the next world, and so perhaps these were graves of tax accountants. However, such deposition in graves reminds me of the joke: "Mike had borrowed and never repaid loans from Pat. At Pat's funeral, Mike decides to pay off his debts, writes a check for $10,000 that more than generously covers his indebtedness, and places it in Pat's casket as it is being lowered into the grave." Thus, an alternate explanation for an ancient burial practice may be that this joke has roots that are at least five thousand years old.

The painstaking and inspired searching through the records of archeological digs and museum inventories that revealed the existence and importance of these counters and the systematic cataloging of them by size/shape/markings was done by Denise Schmandt-Besserat. The main conclusion of her work, published in the 1970s, was an entirely new conception of how writing began with such counters. This is not the interest of this book and so I will not pursue this aspect of her work further, except to praise it.

However, in the 1980s Schmandt-Besserat published her conclusions about the numerical implications of her finds. Her opinion was that the use of counters (she refers to them as tokens), such as those described here in table 3.2.3, marked a transition to the conception of *abstract counting* from a previous conception of *concrete counting* or *archaic counting* that was embodied in the earlier tokens that defined both number and identity. I have already discussed these concepts in section 2.1 and have rejected such a transition. I will therefore just briefly state my interpretation of the token findings.

Early in token evolution, both the number and the identity of what was being counted were defined by the size/shape/markings of the token, but that does not mean nobody was aware of an abstract concept of number not associated with the identity of what was being counted or measured. Later it was realized that it was more efficient to have just one sequence of tokens to define the number of anything, as, for example, the tokens in table 3.2.3, with a separate token added to the set, or the envelope of the set marked, to identify what was counted or measured.

By the end of the pebble-counting era, around 3000 BCE, in both Egypt and Babylon, there was a general ability to count to large numbers, and to add and subtract even large numbers efficiently. When a compatible number and measuring system (base-60) was introduced in Babylon, it was just as logical and good as the Egyptian solution (base-10). But when the number systems were required to perform more complex arithmetic, the difficulties associated with a large base became apparent; base-10 survived, base-60 did not. As we shall see in chapter 5, multiplication and division were so awkward using base-60, we can be sure that base-60 evolved long before there was much need for these more sophisticated arithmetic operations.

Vestigial use of base-60 also survives today in our angular measurement units, because 360 (a rounded-off year) is such a natural, intuitive, convenient division of a full rotation. When the metric system was first imposed in France at the beginning of the nineteenth century, one of the proposals was that a full rotation be divided into 400 rather than 360 degrees. This would have made 1 degree of the earth's circumference equal to 100 km, which would have been very convenient. But the French public would not accept such disruption in the natural order of things, and that units change is not now part of the metric system.

3.3 In the Jungles of the Maya

There will always be an unsatisfying limit to what an archeological record that dates back more than five millennia can reveal. However, another way to gain insight into the *childhood of mathematics* is to look at a completely independent culture that went through its urbanization stage much later, hence from which more evidence might have survived.

The Mayan culture in Central America entered its written-language, written-mathematics urban stage by the first millennium, some three thousand years after Babylon entered that stage. Although in a state of decline by the time Spanish conquerors arrived in the sixteenth century, many Mayan texts still existed that possibly could have been able to give a detailed description of Mayan mathematics. Unfortunately, in their zeal to eradicate the Mayan religion and convert the Maya to Christianity, the Spaniards burned most of the Mayan texts.

Only three major texts survived: the Dresden Codex, the Paris Codex, and the Madrid Codex, so named for the cities where they are now preserved. Only one of these, the Dresden Codex that records astronomy observations, is the source of most of what we know about Mayan mathematics. Probably written around the year 1100, this codex contains seventy-eight pages made from wood-bark paper, and each page is about 8.5 cm × 20.5 cm. That might appear to be a sizable record, but the Mayan hieroglyphic writing does not produce much content per page. Rather than being a more complete record than the Babylonian, the Mayan mathematical record is much less detailed, but sufficient information has survived to make it useful to study. Ironically, a book written by Diego de Landa, the very priest who burned the Mayan texts, is an important reference about Mayan culture.

During an ice age between about forty thousand and ten thousand years ago, hunter-gatherers crossing a land bridge from Siberia to Alaska became the forebears of all Native Americans. What makes the cultures of the Native Americans especially interesting is that for at least ten thousand years prior to the discovery of America by Columbus, cultural progress in the Americas was apparently independent of developments in Europe and Asia because the Siberia-Alaska land bridge vanished under rising seas at the end of the ice age. There had been Viking intrusions into northeastern America around the year 1000, and there may have been occasional intrusions from across the Pacific, but at least to date, these intrusions do not appear to have significantly affected developments in the Americas.

Like the Celts of Europe and many other cultures in a hunter-gatherer stage, Mayan counting probably started using fingers and toes, and eventually evolved into a base-20, or *vigesimal*, written number system. Like the Babylonian, this is also a large-base system. As illustrated in table 3.3.1, the symbols of the base are *additive*, and as in Babylonian cuneiform, just two symbols are used.

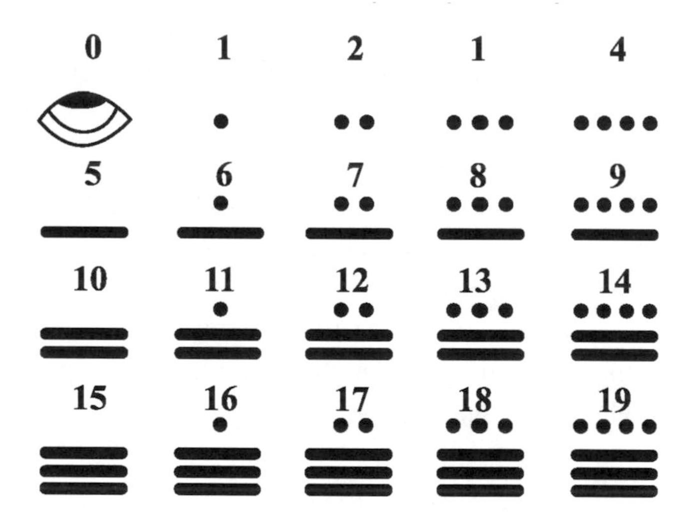

Table 3.3.1 *Additive Mayan symbols for numbers 1 through 19*

There is no archeological record from a prewriting stage, but the written symbols suggest that they possibly evolved from the use of *counters*: a pebble for unity and a stick for five. However, it is also possible that five pebble-symbols ●●●●● simply evolved into ▬▬▬ with the invention of writing in accordance with the "rule of four" deemed previously relevant to figure 3.2.1. Mayan base-20 numbers can also be accounted for as descendants of finger counting described in section 1.2: each finger on the right hand has a value of 1; each finger on the left hand has a replacement value of 5. Mayan counting is physiologically based on hands and feet, as was counting by many other Native American cultures (see section 2.2).

Just as the Babylonians did, the Maya also invented a *positional* system; but as the Babylonians did not do, the Maya invented a zero symbol. Table 3.3.2 is an example of a *positional* Mayan number and its decimal evaluation. I have transcribed the Mayan numbers into decimal symbols using the same notation I use for any number system with a base greater than 10: colons separate positions (see section 1.3). Because of the use of the zero, and the fact that the Maya apparently only knew how to write integers, there is no ambiguity in the value of Mayan numbers, which we have seen was a deficiency of Babylonian numbers. Conversion of a Mayan number to a decimal value uses equation (1.3.1).

$= 1 \times 20^4 = 1 \times 160{,}000 = 160{,}000$

$= 6 \times 20^3 = 6 \times 8{,}000 = 48{,}000$

$= 0 \times 20^2 = 0 \times 400 = 0$

$= 12 \times 20 = 240$

$= 15$

$1{:}6{:}0{:}12{:}15 = 160{,}000 + 48{,}000 + 0 + 240 + 15 = 208{,}255$

Table 3.3.2 *Example of a Mayan number for counting things*

The Maya sometimes wrote their numbers with the positions arrayed vertically, but they also sometimes arrayed them horizontally, just like cuneiform numbers:

1:6:0:12:15 =

Figure 3.3.1 is a page from the Dresden Codex. The gruesome figures are Mayan hieroglyphs, or glyphs for short. For many years, the glyphs were frustratingly untranslatable. But since the 1960s there has been amazing progress, and about 90 percent of the Mayan text is now translatable. The number symbols are readily differentiated from the glyphs and are clearly readable in their vertical and horizontal arrangements. Deciphering of the numbers is largely independent of the ability to translate the glyphs, and they were deciphered at the beginning of the twentieth century, long before the Mayan language was translated. In fact, understanding the numbers has been helpful in translating the glyphs.

Prior to the invention of positional numbering, the Maya, like the Babylonians, had an *additive* number system with different glyphs for different powers of 20. Thus, both cultures appear to have gone through very similar mathematical evolutions: from finger and spoken counting in a hunter-gatherer stage, to pebble counting in a herder-farmer stage, to written additive numbers, and then finally to written positional numbers in an urban stage. Considering the fact that there was absolutely no contact between Babylonians and the Maya, this suggests the universal sequence of evolution of number systems proposed previously in section 1.2. Most societies do not go through all these stages because at some point they adopt the superior technique of a neighboring or conquering culture.

We know essentially nothing about how the Maya did arith-

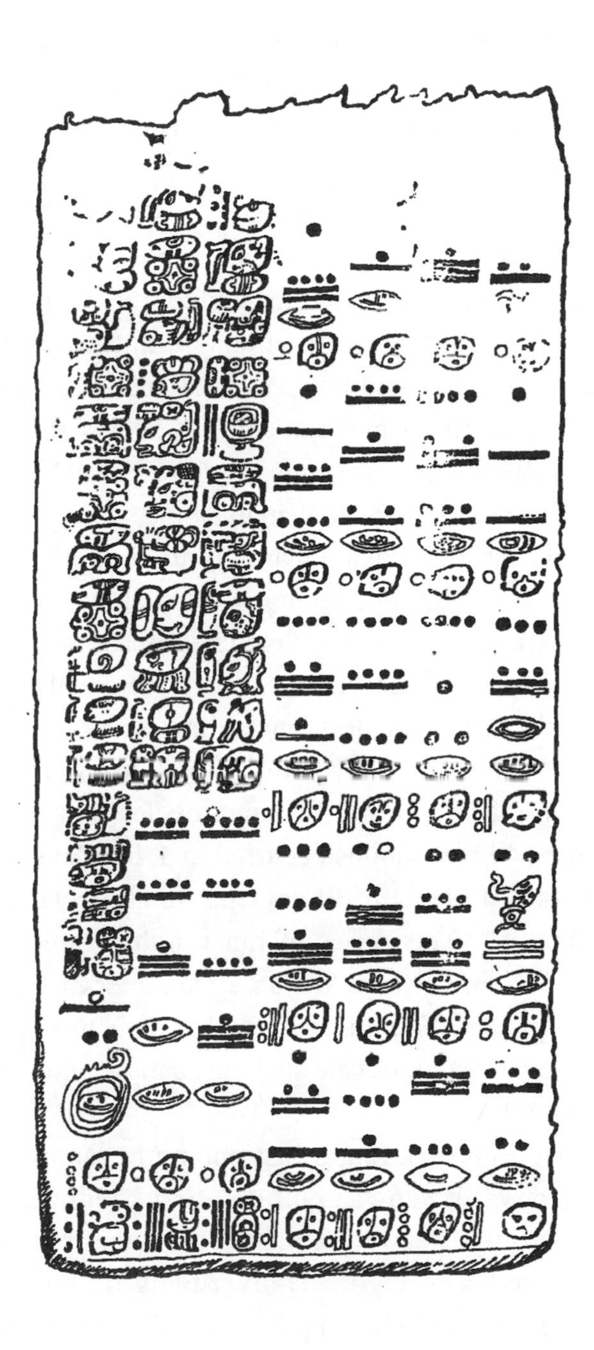

Figure 3.3.1 *A page from the Dresden Codex*

metic or of what mathematics they were capable. They must have had some arithmetic capability, for there surely were calculations required for administrative purposes—taxes are one of the two unavoidables.

The surviving Mayan mathematical records deal only with astronomy observations and the calendars based on them. No record of the transactions of ordinary life remains. In fact, the use of pure base-20 counting of just plain "things," as illustrated in table 3.3.2, is only justified by records of such use by the Aztecs. All the various Central American cultures (Maya, Aztec, Zapotec, and Mixtec) used base-20 systems. Attributing such counting also to the Maya, even though no Mayan example survives, appears reasonable. However, the Maya apparently used another, but closely related, number system for astronomy calculations.

Mayan astronomers did not have telescopes; they did not have an angular measure concept; they did not have metals; and they did not have clocks. Their smallest unit of time was *one day*, and their observations were essentially limited to counting the number of days that elapsed for various planetary cycles. For example, they observed that 149 new moons occurred in 4,400 days. If we do the arithmetic, 4400/149 = 29.530 days per lunar cycle. The modern value is 29.53059. This series of lunar sightings encompasses a twelve-year period. Clearly, Mayan astronomers were amazingly accurate and patient. You might think that their lives depended on these measurements to motivate such dedication, and that is exactly what they thought.

Mayan astronomy began, as astronomy began everywhere, to serve astrology. When a culture first begins to employ astronomy/astrology, the initial results are wondrous indeed. The Maya worshiped a sun god. Daily rituals of human sacrifice induced the sun to rise. Mayan priests consistently produced good results.

It is a mystery to me how a human, who all his life had observed the sun to rise without fail each day, was convinced by priests that their ritual was required. It was not unique among the Maya. There were similar rituals in ancient Egypt and among many other cultures worldwide. Eclipses may have been frightening events, but they hardly last enough time for priests to robe-up for their hocus-pocus. Perhaps such a ritual originated when some huge volcanic eruption or meteor impact blotted out the sun for at least a few days. That would surely have incited priests to an orgy of ritual that would eventually appear to have induced the sun to return.

The yearly solar cycle is also easy to determine by simple observations of the position of the rising sun, or its position at high noon, or the star positions in the night sky. This defines a useful sequence of events:

> A time to be born, and a time to die; a time to sow and a time to reap;
> A time to kill, and a time to heal; a time to break down and a time to build up;
> A time to weep, and a time to laugh; a time to mourn, and a time to dance . . .
>
> Ecclesiastes (Old Testament); Qoheleth (Hebrew Bible)

With such successes, Mayan astrologer/astronomer priests justifiably gained prestigious and powerful roles.

The planets are very prominent in the night sky, and dramatic events, such as the birth of a king or a disastrous drought, sometimes coincide with exceptional planetary occurrences, such as Saturn and Jupiter almost touching, or the moon eclipsing Venus. It is thus understandable that people came to believe that the planets also affect our lives. However, planets move in mysterious ways when observed from earth, and it is not possible by casual observation to predict the motions of the planets. Mayan astrologer/

astronomer priests assumed the task of determining planetary occurrences, thereby predicting earthly events.

Nowadays we are forgiving when the rituals and the prayers of our priests fail to produce their intended divine intervention and the sick still die and the rains still do not come. We are forgiving because the cost of this service is so inexpensive. Mayan astrologer/astronomer priests demanded much greater payment—an orgy of human sacrifice. That their priests failed to produce results after exacting such a price might have been the reason why the Maya abandoned their ceremonial cities and faded back into the jungle.

In recent centuries, there has certainly been a worldwide tendency to migrate to cities, but from Neolithic times onward, many cycles of migration both to and from cities have occurred throughout the world. One particularly well-understood migration might be relevant to the Mayan abandonment of their cities. Anthropological studies of settlement history in the Middle East in the nineteenth and twentieth centuries show that onerous taxation and the threat of conscription into the Ottoman army drove many villagers to abandon their homes in agricultural regions and disappear into the desert as nomad-herders, a more secure, if less comfortable, life.

Actually, nobody knows why the Mayan empire disintegrated. Some think it is because too intensive farming led to an ecological disaster. This explanation is currently popular because well-meaning archeologists are trying to teach a lesson: be good to your environment, or it will kill you. Archeologists sometimes feel compelled to justify their "useless" occupation. To me the ecological explanation makes little sense. A decline in agricultural output can require a decline in urban population, but not abandonment of ceremonial centers.

Are archeologists ever guilty of allowing their personal politics or prejudices to influence their perceptions? Of course they are, and their initial, completely mistaken image of the nature of Mayan culture is a glaring example. Up until the middle of the twentieth century, they portrayed the Maya as peaceful, gentle, and rich in artistic and intellectual accomplishments. When eventually the Mayan glyphs were translated, this idealized picture crumbled. The Maya had been as typically aggressive, savagely cruel, and human sacrificing as all people were at this stage of development, and many were and still are at much later stages.

Interestingly, the idyllic perception of the Maya, particularly by the English archeologist Eric Thompson (1898–1975), was a rerun of the idyllic perception of Minoan culture by Arthur Evans (1851–1941), the English archeologist who had excavated the palace at Knossos, Crete, some forty years earlier. Archeologists, like the rest of us, sometimes see what they want to see. Was Thompson simply copying Evans's vision? My guess is that both, probably without even realizing it, were simply expressing the generally accepted, ancient, biblical vision that there had once been a Garden of Eden, an uncorrupted, pristine world. Evans and Thompson apparently both believed that they had dug back far enough and had discovered that past. Presumably, these well-intended archeologists wanted to demonstrate that the grime and crime of Victorian England was not inevitable and that humans were capable of creating a much better world. Unfortunately, no matter how deep we dig, no such history will be found. Survival for millions of years in a harsh world has required the evolution of aggressive, acquisitive, and promiscuous instincts. Utopias are not in our past; and alas, they are probably not in our future either because when the lion lies down with the lamb, the lion will soon starve to death and become extinct, contrary to the biblical prophesy of Isaiah (Old Testament), or Yeshayahu (Hebrew Bible).

* * * * *

The Mayan number system for astronomy was deciphered by comparing astronomy observations in the Dresden Codex with modern values. The following series of observations was recorded in the Dresden Codex: [8:2:0], [16:4:0], [1:4:6:0]. Assuming that these numbers represent counting the number of days of some astronomical cycle and applying pure base-20 position values produced these results:

$$[8:2:0] = (8 \times 20^2) + (2 \times 20) + (0 \times 1) = 3{,}240 \text{ days}$$
$$[16:4:0] = (16 \times 20^2) + (4 \times 20) + (0 \times 1) = 7{,}480 \text{ days}$$
$$[1:4:6:0] = (1 \times 20^3) + (4 \times 20^2) + (6 \times 20) + (0 \times 1) = 9{,}720 \text{ days}$$

Figure 3.3.1 records the Venus observations. The Mayan number written in vertical alignment in the lower right-hand corner can readily be read as 8:2:0. Next to it is clearly 16:4:0, and the third entry is clearly 1:4:6:0; however, the fourth number has a copying error.

These counts do not correspond to presently known cycles of any of the planets. However, slightly modifying the interpretation of the Mayan numbers produced the results:

$$[8:2:0] = 8(18 \times 20) + (2 \times 20) + 0 = 2{,}920 = 5 \times 584 \text{ days}$$
$$[16:4:0] = 16(18 \times 20) + (4 \times 20) + 0 = 2 \times 2{,}920 = 10 \times 584 \text{ days}$$
$$[1:4:6:0] = 1(18 \times 20) \times 20 + 4(18 \times 20) + (6 \times 20) + 0 = 3 \times 2{,}920 = 15 \times 584 \text{ days}$$

The cycle time, called the *synodic period* in astronomy jargon, for the planet Venus is 584 days. Thus was Mayan numbering in astronomy deciphered. The generally accepted interpretation of these data is that they represent counting with a deviate base-20 system. This interpretation does not conform to the universal and natural tendency to use just one replacement number for counting things. My interpretation is that Mayan recording of synodic periods may appear to be *counting*, but it is really *measuring* time. As we have now seen from numerous examples (for example, in table 3.2.6), measuremnts based on natural units of convenience do not have just one replacement number.

The Mayans *defined* a *year* as 360 days, 18 months of 20 days each. They knew that a real year was about 365 days, and a real year was a defined year plus five evil days. Pity the poor Maya who were born during the five evil days because they were a cursed lot. The Egyptians, Babylonians, and Maya all had a defined year of 360 days: the Egyptians because $360 = 12 \times 30$, the Babylonians because $360 = 6 \times 60$, and the Maya because $360 = 18 \times 20$. Again, we have an example of natural and intuitive definition of a unit of measure in nicely rounded, even numbers, with many useful divisors.

To me it is obvious that the Maya defined time intervals longer than one year just as we do nowadays, by using *mixed units* of years and days. Thus, the Venus cycle times should be more properly translated, not by replacing 20^2 with 18×20, but by replacing 18×20 with the word *years*, and with an option of referring to the second position as months although this option does not affect the total count of days:

[8:2:0] = 8 years + 2 months + 0 days = 8 years + 40 days
= 2,920 days

[16:4:0] = 16 years + 4 months + 0 days = 16 years + 80 days
= 2 × 2,920 days

[1:4:6:0] = 24 years + 6 months + 0 days = 24 years + 120 days
= 3 × 2,920 days

(Can you now identify the copying error in the fourth number in figure 3.3.1?)

Reading from right to left, the first two positions measure time in days and the position values are in a perfect base-20 sequence; the remaining positions measure time in years and are in a perfect base-20 sequence. The first two positions are capable of counting to 399 (= [19 × 20] + 19), but in astronomy are never used to count to greater than 359 (= [17 × 20] + 19) because 360 days is then replaced by one defined year.

FUN QUESTION 3.3.1: The synodic period of Mars is 780 days. How would the Maya have written this? Also, give the answer in decimally transcribed Mayan numbers.

Apparently, the difference between counting and measuring was too subtle to be appreciated by the curator of the Dresden Museum who analyzed this codex at the end of the nineteenth century. But he was also the one who brilliantly recognized the recording of the Venus cycle, so we can forgive him his oversight in not differentiating between counting and measuring. Unfortunately, generations of archeologists and mathematicians have uncritically quoted his mistake. Repetition has made the conclusion that the Maya used a deviate base-20 number system for astronomy into gospel, but like much gospel it also ain't necessarily so.

As already noted in sections 3.1 and 3.2, we use mixed units because they are more easily visualized. It is only necessary to convert from mixed units to a single unit for doing some calculations. Since the Maya apparently did not do much if any calculating that required such conversion with their astronomy observations, mixed units was not a problem. Babylonian, base-60 use for everyday recording of time similarly survives because we seldom have to do calculations that require conversion to a single unit. But when it is necessary, it is a nuisance.

FUN QUESTION 3.3.2: Calculate the number of seconds in a 365-day year.

Unfortunately, for decipherers of the Dresden Codex, the Maya did not invent a symbol to separate the year count from the day count; neither did they bother to note the units because they were obvious to them. In present-day use of common mixed units, we also frequently do not bother to note the units involved. Have you ever heard this song from the 1930s?

Five foot two,
Eyes of blue,
Oh, how she could kootchy-koo,
Has anybody seen my gal?

It is obvious to anyone familiar with English units that "my gal" is 5 feet, 2 inches tall. In fact, in common usage this would still be obvious even if all that was said was "my gal" is five two, but that would not have been consistent with the rhythm of the song.

* * * * *

The construction date inscribed on a monument in the ancient Mayan city of Tikal is [8:12:14:8:0]. This yields a Mayan date of $([8 \times 20^2] + [12 \times 20] + 14)$ years + $([8 \times 20] + [0 \times 1])$ days = 3,454 years + 160 days. Day zero, the Mayan date for the creation of the world, is frequently given as 3113 BCE. Finding the relationship between the Mayan calendar and our present calendar is known as the *correlation problem*. There is some uncertainty about this Mayan creation date, but we shall not concern ourselves here with the details of the correlation problem. Using this creation date, we can convert the construction date of the Tikal monument into a date according to our calendar: 3,454 – 3,113 = 341. The glory days of the Mayan empire were during the first millennium, and so this is certainly a reasonable date.

Rabbis of the first century deduced from the Hebrew Bible that the date of creation was 3760 BCE. Considering the fact that both the Mayan and the Jewish creation dates miss the true creation date by about fourteen billion years, the closeness of these two accounts to each other is remarkable. Is this just a coincidence, or is there some logic to this?

The concept of evolution implies universality: similar conditions tend to produce similar results. What does an arid desert and a tropical jungle have in common that two such radically different environments motivated the Babylonians and the Maya to invent similar counting systems over comparable historical intervals? The answer is clearly not similarities in physical conditions, but rather universality of brain function.

We should also not be surprised that religious evolution among the Jews of Canaan and the Maya was similar. At first, it might appear that the religions were radically different, with the Mayans engaging in savage human sacrifice while the Jewish religion is devoted to benign study of the Torah. But that is comparing the Mayan religion then with the Jewish religion now. The Jewish religion only abandoned animal sacrifice after the Romans destroyed the temple in Jerusalem in the year 70. One of the most famous of biblical stories, Abraham's aborted sacrifice of his son Isaac, is a reminder that human sacrifice had been practiced in past centuries. Jewish Canaanite and Mayan religions had much in common at comparable periods in their evolution.

Religion does not start with such profound questions as when and how the universe was created. Rather it starts with questions about what death is, and where do dead ancestors go? For every religion, ancestor histories, real and invented, provide the measure for how far back in time history extends. Logic demands that the creation myth precede all other myths in presentation sequence, even though it was not the first myth invented. Much later, after a culture has developed an ability to do arithmetic, a creation date may be calculated. The fact that the Mayan and Jewish creation dates are so similar reflects a similarly remembered duration of ancestor history, and that the years when the calculations were made were not far apart. Now that Mayan glyphs can be read, translations of Mayan inscriptions on monuments and buildings reveal that the Maya were indeed dedicated recorders of genealogies, just as the Jews in Canaan were.

Now that we understand Mayan base-20 counting, let us compare it with the other large-base system, the Babylonian base-60. The Babylonian *counters* replacement scheme was an alternating 1-for-10 and 1-for-6 sequence to yield a base-60 system. The Mayan replacement scheme was an alternating 1-for-5 and 1-for-4 sequence to yield a base-20 system. Very similar systems, but arrived at by different evolutionary paths. Like the almost universally adopted base-10, Mayan base-20 is also simply explained by physiology. On the other hand, the Babylonian base-60 resulted from an interaction between measuring units and counting (see section 3.2).

We know essentially nothing about Mayan arithmetic, but we can certainly assume that at a minimum they knew how to add. A base-20 addition table has 190 entries (see table 1.3.4), so it is unlikely that they used memorization of an addition table. Babylonian and Mayan number systems arrived at their respective large bases via different evolutionary paths, but their structural similarity means that they probably used similar arithmetic methods. A reasonable guess is thus that the Maya did similar abacus addition. Table 3.3.3 illustrates a possible Mayan abacus addition modeled after the Babylonian abacus calculation previously given in table 3.2.2. Like Babylonian symbols, Mayan symbols can also be described as *abacus notation*, and we thus can be reasonably confident that they also did abacus addition.

Position value	20^2		20		1	
	x5	x1	x5	x1	x5	x1
Replacements		●		●	▬	
Addend = 17:14 = 354_{10}			▬▬▬	●●	▬▬	●●●●
Addend = 15:8 = 308_{10}			▬▬▬		▬	●●●
Sum = 1:13:2 = 662_{10}		●	▬▬	●●●		●●

Table 3.3.3 *Abacus, Mayan vigesimal system addition*

How, or even if, the Maya did any other arithmetic is unfortunately unknown. It would have been invaluable to know how other Mayan arithmetic compared with Babylonian arithmetic. Damn you Diego de Landa, would that your imagined torments of hell really existed for you to suffer eternally for your sanctimonious vandalism.

Even had the Spaniards not arrived, Mayan progress in mathematics would have aborted. Nevertheless, no other Native Americans produced such a sophisticated number system. Despite its small size, the surviving Mayan record has been critical for appreciating the universality of evolution, particularity the universality of mathematical intuition.

4

MATHEMATICS IN THE VALLEY OF THE NILE

4.1 Egyptian Multiplication

The principal source for knowledge of ancient Egyptian mathematics is the Rhind Mathematical Papyrus, now held in the British Museum. The Rhind Papyrus is a scroll 5.5 m long and 0.33 m wide. It contains a table of fractions and eighty-seven mathematical problems. The papyrus, written in about 1700 BCE, was found among the ruins of ancient Thebes in Egypt. Henry Rhind, a Scottish collector of antiquities, purchased it in 1858. The writer of the papyrus was Ahmes the Scribe, and so this papyrus is sometimes referred to as the Ahmes Papyrus. Ahmes wrote that he copied it from a papyrus written in about 1800 BCE. It was presumably a guide for teaching mathematics to apprentice scribes and therefore probably related to standard, advanced material going back to even earlier times. To how much earlier is not known.

The Rhind Papyrus records how the Egyptians multiplied; we do not multiply that way nowadays. Our present method of pencil/paper multiplication is to memorize the thirty-six entries in the multiplication table given in table 4.1.1. The number of entries can either be tediously counted from the table or calculated using equation (1.3.2) for the sum of an arithmetic series ($N = 8$, $a_{AVG} = 4.5$, $S = 8 \times 4.5 = 36$). Three simple and obvious rules eliminate the need to memorize more products: $m \times n = n \times m$, $0 \times m = 0$, and $1 \times m = m$, where m and n are any numbers.

x	2	3	4	5	6	7	8	9
2	4	6	8	10	12	14	16	18
3		9	12	15	18	21	24	27
4			16	20	24	28	32	36
5				25	30	35	40	45
6					36	42	48	54
7						49	56	63
8							64	72
9								81

Table 4.1.1 *Multiplication table—decimal system*

Memorization of this thirty-six-entry table is no trivial matter; we devote a few years of daily practice in school before more or less mastering the multiplication table. Indeed, in our age of the ubiquitous electronic calculator, pencil/paper multiplication of any but the simplest products is almost a lost art. Early in the history of European base-10 use, memorization of the complete, thirty-six-entry multiplication table was considered too onerous, and various aids to multiplication were employed. An interesting example is the method of Robert Record, a Welsh mathematician. In his textbook from 1542, he taught how to manage with memorizing only the ten, left-most entries of table 4.1.1. The derivation of Robert Record's method is interesting. The product of any two integers, m and n, can be written as a two-digit number: $mn = 10t + u$, where t = tens digit and u = units digit.

Rewriting m as $[10 - (10 - m)]$ and n as $[10 - (10 - n)]$, $mn = [10 - (10 - m)][10 - (10 - n)] = 10(m + n - 10) + (10 - m)(10 - n)$.

Therefore $t = (m + n - 10) = m - (10 - n) = n - (10 - m)$ and $u = (10 - m)(10 - n)$.

Robert Record showed how to remember this calculation with the mnemonic diagram shown in figure 4.1.1 that also illustrates a sample multiplication of 6×8:

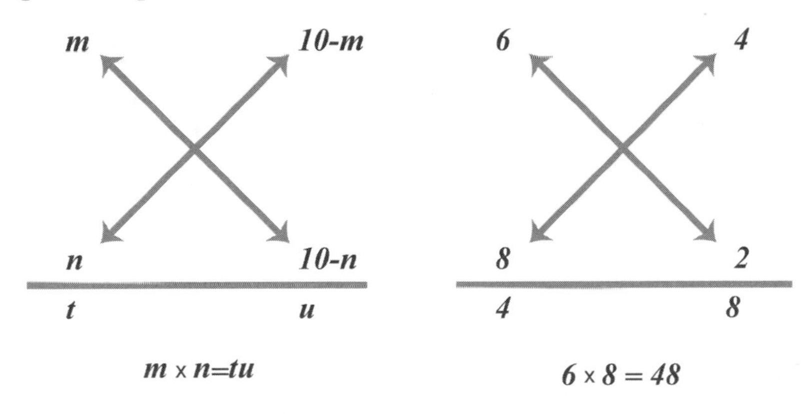

Figure 4.1.1 *Multiplication mnemonic*

The difference between numbers at the ends of the diagonals gives the tens digit; the product of the terms on the right-hand side gives the units digit.

Using the Egyptian's *additive*, base-10, hieroglyphic system, multiplication using the multiplication-table method would have required a thirty-six-entry table for each combination of symbols (units × units, units × tens, and so on). However, each thirty-six-entry table relates to every other table by a simple shift in symbols, so multiplication by memorization would have been just a little more difficult than nowadays. Table 4.1.2 compares modern and hypothetical *hieroglyphic* multiplication using this multiplication-table method. For multiplication with *hieratic* symbols, the number of products that it is necessary to memorize is the same, but there are of course nine times as many symbols to memorize, but that is an independent problem.

$$\begin{array}{r} 4 \\ \times\ 3 \\ \hline 12 \end{array} \qquad \begin{array}{r} 40 \\ \times\ 3 \\ \hline 120 \end{array} = \begin{array}{r} 4 \\ \times\ 30 \\ \hline 120 \end{array} \qquad \begin{array}{r} 40 \\ \times\ 30 \\ \hline 1{,}200 \end{array} \qquad \begin{array}{r} 44 \\ \times\ 33 \\ \hline 1{,}452 \end{array}$$

Table 4.1.2 Hieroglyphic multiplication-table method of multiplication

Note how a simple shift in symbols relates the products 3×4, 3×40, and 30×40.

My guess is that any Egyptian scribe who daily performed many multiplications probably did remember many frequently occurring products. However, considering the memorization burden of the multiplication-table method, it is understandable that the Egyptians invented another method that required much less memorization, and which is now called *Egyptian multiplication*. They performed multiplication by a sequence of two-number additions, a *multiplication algorithm*.

Most Egyptians and Babylonians in this era were illiterate, with reading/writing limited to a specially educated class of scribes. As discussed in section 3.1, intuitive, pebble-counting arithmetic and abacus arithmetic had been widely used in Egypt and Babylon even prior to the invention of writing. It is illogical that the general population lost this ability after the invention of writing around 3000 BCE. Presumably, written arithmetic was limited to the scribe class, and it is only this method, as presented in the Rhind Papyrus, that I shall discuss.

Let us suppose that an Egyptian scribe wished to multiply 13×173. He could simply do the addition of the two numbers $173 + 173 = 346$, and then do the addition of the two numbers $173 + 346 = 519$, and so on, each time adding 173 to the partial sum for twelve successive additions. Presumably, this is the way it was done until some lazy scribe invented a shortcut. He added each partial sum to itself, thereby doubling the partial sum with each addition, rather than just increasing the partial sum by the value of the addend. Then he added up the relevant partial sums. Again using the multiplication 13×173 as an example, he did the sequence of additions:

$$
\begin{array}{rcll}
173 & = & 1 \times 173 & \\
\underline{+\,173} & & & \\
346 & = & 2 \times 173 & 1\times + 1\times = 2\times \\
\underline{+\,346} & & & \\
692 & = & 4 \times 173 & 2\times + 2\times = 4\times \\
\underline{+\,692} & & & \\
1{,}384 & = & 8 \times 173 & 4\times + 4\times = 8\times
\end{array}
$$

Here doubling stops because $8\times + 8\times > 13\times$.

$$
\begin{array}{rcll}
\underline{+\,692} & & & \\
2{,}076 & = & 12 \times 173 & 8\times + 4\times = 12\times \\
\underline{+173} & & & \\
\mathbf{2{,}249} & \mathbf{=} & \mathbf{13 \times 173} & \mathbf{8\times + 4\times + 1\times = 13\times}\ (1101_2 = 13_{10})
\end{array}
$$

This calculation may look familiar because it is just the *greedy algorithm* used previously (see FUN QUESTION 1.3.3) to convert a decimal number into a binary number. No base-10 multiplication table must be memorized, but in its stead just the much more easily memorized sequence of powers of two, 2^n (2, 4, 8, etc.).

In the Rhind Papyrus this multiplication algorithm appears in an abbreviated format:

\	173
	346
\	692
\	1,384
Total	2,249

The mark "\" denotes the numbers to be added together to complete the multiplication. Either the scribe did the additions on a scratch paper, or he used an abacus. No surviving document shows the details of how the Egyptians did addition.

If performed by the multiplication-table method, 13×173 requires six multiplications (3×3, 3×7, 3×1, 1×3, 1×7, 1×1) plus one addition for a total of seven operations. Only five operations (count the + signs) are required using Egyptian multiplication. Of course, the Egyptians did not use Hindu-Arabic symbols as in this example, but the operations are valid for whatever symbols for numbers are used.

The ancient Egyptians clearly understood that a combination of powers of two could define any integer. In other words, they understood the operation of the binary system (see section 1.3). This may have been the first ever use of the base-2 system; at least it is the first recorded use. (See section 2.2 for earlier but doubtfully valid evidence.)

For *additive* Egyptian numbers, their method of multiplication may have been even more efficient than the modern, multiplication-table method. For those who do arithmetic infrequently, it is easy to forget a once-memorized multiplication table. For cultures without a formal education system, learning a multiplication table is difficult. For such people Egyptian multiplication is fine and has been used through recent times as so-called *peasant multiplication*.

I recently observed Egyptian multiplication used at a blackjack table. Someone asked the dealer for 20 chips. Rather than just count out 20 chips, the dealer counted out a stack of 5. Then he measured out a stack of the same height; he now had $2 \times 5 = 10$ chips. Now he placed one stack on the other and measured out another stack of the same height; he now had his $4 \times 5 = 20$ chips. Of course, this was all done with blinding speed, notwithstanding dramatic hand flourishes. Egyptian multiplication survives on the gaming tables of Las Vegas—coincidentally home also of a large pyramid.

FUN QUESTION 4.1.1: How many additions are required to calculate 137^2 by Egyptian multiplication?

Consider the division $x = b/a = 2{,}249/173$. Division can be thought of as how many successive subtractions of a from b are required such that $b - xa = 0$. To calculate x, we can perform successive subtractions: $2{,}249 - 173 = 2{,}076$; $2{,}076 - 173 = 1{,}903 \ldots 173 - 173 = 0$ and find that we had performed 13 subtractions, so $x = 13$. However, division can also be thought of as how many successive additions of a are required to obtain b, or in algebraic notation as $xa = b$.

Egyptian division used multiplication by successive additions of two numbers, but they used their multiplication shortcut. Thus, to do the division 2,249/137, the scribe would have performed the six-step *division algorithm*:

1×	173			
	+173			
2×	346			
	+346			
4×	692			
	+692	⟶	692	
8×	1,384	⟶	+1,384	
	+1,384		1,976	4×+8× =12×
16×	2,768		+173	
			2,249	**1×+ 4×+ 8× = 13×**

When it is seen that 16× is too greedy, it is then clear that the correct result is 8× plus some combination of smaller greedy multiples, and these are easily found to be 4× + 1×.

I can also envision another way that the use of the binary system may have naturally evolved. A merchant in an Egyptian market sells his product by weight, which he weighs for each customer on an equal-arm balance as illustrated in figure 4.1.2.

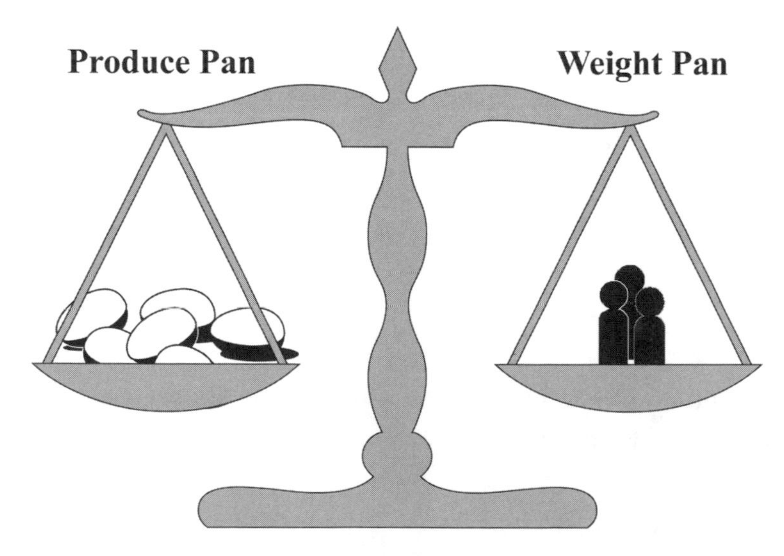

Figure 4.1.2 Equal-arm balance

He sells his product in increments of one kite. This was an ancient Egyptian weight measure, equal to about 30 grams. He keeps a collection of 1-kite weights. However, weighing out many kite purchases is tedious with only 1-kite weights; and anyway, the small 1-kite weights tend to get lost. Therefore, he decides to replace some of his 1-kite weights with larger weights. After trying different weight combinations, he, as you now must realize, concludes that a set of weights of 1, 2, 4, 8 . . . kites, provides a weight set with the minimum number of weights. This is how the binary system may have been unknowingly invented in a marketplace in ancient Egypt.

FUN QUESTION 4.1.2: A merchant, who was the great-great-grandson of the merchant who invented the binary weight-set system (change was not rapid in ancient Egypt), realized that it was not necessary to only put weights in the weight pan. He realized that he could also put weights in the produce pan, which could then be subtracted from the weights in the weight pan. What is the weight set with the minimum number of weights required to measure in increments of 1 kite, when weights can be added to either pan?

ANSWER:	weight pan	produce pan	net weight
	1	0	1
	3	1	2
	3	0	3
	3 + 1	0	4
	9	3 + 1	5

Continue this table to show that the minimum weight set is: 1, 3, 9, 27 . . . (3^0, 3^1, 3^2, 3^3 . . .) kites.

These are the position values of the base-3, or *ternary*, system! Thus, the ternary system might also have been invented in a marketplace in ancient Egypt. Although this system required far fewer weights than the binary system, the merchant found the arithmetic confusing, and he invented a new system that also used weights in both pans. Although it did not provide a weight set with the absolute minimum number of weights, he could now more easily use base-10 arithmetic with weights 1, 2, 5, 10, 20, 50, 100, 200, 500 . . . kites. This is essentially the weight-set sequence still used almost universally with manual equal-arm balances.

Weighing with an equal-arm balance and defining currency denominations have much in common. Thus, it is not too surprising that this same 1, 2, 5 . . . sequence is at least approximately followed by most decimalized currencies. It is exactly the sequence of denominations used for the euro, a recently issued currency. For American currency, there are some minor deviations: a 2¢ coin is missing, and rather than a 20¢ coin there is a 25¢ coin, but dollar denominations follow the sequence exactly. However, when we consider the denomination sequence in terms of replacement numbers, we see that both the euro and the US denominations use the same method. To almost obtain the natural and intuitive sequence of doubled units (base-2), while making calculation easy by making the sequence compatible with a base-10 arithmetic, both currencies use a replacement sequence of two 2s and one 2.5 ($2 \times 2 \times 2.5 = 10$). Figure 4.1.3 illustrates the replacement sequence for the euro. (See section 4.2 for more examples of natural and intuitive use of a sequence of doubled units.)

Denominations	1		2		5		10		20		50		100
Replacements		2		2.5		2		2		2.5		2	

Table 4.1.3 *Replacement sequence for the euro*

In the Indus Valley, there are extensive remains of cities that had populations in the tens of thousands, dating back to 3000 BCE.

This culture is known as Harappan, taking the name from the major city of Harappa. The building material was largely mud brick and, because of the heavy seasonal rains (monsoons) in the region, there is not much of an archeological record. Neither has their script been deciphered yet, so not much is known about their mathematics. Nonetheless, archeologists have found Harappan scale weights with the sequence: 1, 2, 5, 10, 20, 50, 100, 200, 500, 1,000, 2,000, 5,000. This shows that they were using a decimal system, which is not very remarkable. What is remarkable is that in five thousand years nobody has found a better solution to the problem of what is the optimum choice of scale weights or currency denominations for users of a base-10 number system.

The Egyptian solution to multiplication was quite a good one, and thus the question arises of whether other contemporary cultures also adopted it. We shall latter see that the Babylonians probably did not, although it would have been a better solution than any they found. Canaanites may have used it. Hebrew uses the same word *kfl* for both *doubling* and *multiplication*. Not decisive evidence, but intriguing. However, the best evidence is from the biblical story of Noah. As the animals left the arc, Noah told them to go forth and multiply. After a while, Noah happened upon two snakes basking in the sun. "Why aren't you multiplying?" Noah asked. The snakes replied, "We can't, we're adders."

4.2 Egyptian Fractions

Probably the most difficult concept we learn in basic arithmetic is fractions. But we can consider ourselves lucky that we do not do fractions in the manner of ancient Egyptians. With the exception of the fraction 2/3, for which there was a special symbol, every fraction was expressed as a series of *unit fractions*, fractions having 1 as the numerator. Table 4.2.1 is a transcription of a table from the Rhind Papyrus for converting fractions of the form $2/n$ into unit fractions.

$$2/n = 1/a + 1/b + 1/c + 1/d$$

n	a	b	c	d	n	a	b	c	d
3					53	30	318	795	
5	3	15			55	30	330		
7	4	28			57	38	114		
9	6	18			59	36	236	531	
11	6	66			61	40	244	488	610
13	8	52	104		63	42	126		
15	10	30			65	39	195		
17	12	51	68		67	40	335	536	
19	12	76	114		69	46	138		
21	14	42			71	40	568	710	
23	12	276			73	60	219	292	365
25	15	75			75	50	150		
27	18	54			77	44	308		
29	24	58	174	232	79	60	237	316	790
31	20	124	155		81	54	162		
33	22	66			83	60	332	415	498
35	30	42			85	51	255		
37	24	111	296		87	58	174		
39	26	78			89	60	356	534	890
41	24	246	328		91	70	130		
43	42	86	129	301	93	62	186		
45	30	90			95	60	380	570	
47	30	141	470		97	56	679	776	
49	28	196			99	66	198		
51	34	102			101	101	202	303	606

***Table 4.2.1** Rhind Mathematical Papyrus—unit fractions*

The algebraic notation in terms of n, a, b, c, and d is mine; the Egyptians and the Babylonians had no such notation. Neither did they have operational symbols (+, –, ×, /).

This table comprises about a third of the Rhind Papyrus, so it was obviously important. It was a useful table for performing additions. Suppose the addition of $2/17 + 2/85$ were required. Using the $2/n$ table, $2/17 + 2/85 = (1/12 + 1/51 + 1/68) + (1/51 + 1/255) = 1/12 + 1/68 + 1/255 + 2/51$. Using the table again to convert the term 2/51: 2/17 + 2/85 = 1/12 + 1/34 + 1/68 + 1/102 + 1/255. With the aid of the table, a series of unit fractions is again obtained after an addition, and any subsequent arithmetic operations only have to operate with unit fractions.

FUN QUESTION 4.2.1: Express 3/7, 4/7, and 4/5 as unit fractions. Use the 2/*n* table.

Although a cumbersome system, *Egyptian fractions*, as they are commonly called, are easy to use and have some useful properties. They were widely used until at least the thirteenth century, long after the demise of the Egyptian empire.

In math jargon, a *common fraction* has the form *p*/*q*, where *p* and *q* are integers, and defines a *rational number*. Rational numbers were the only kind known to the ancient Egyptians (see section 6.3 for more about *irrational numbers*). Table 4.2.2 shows that somc common fractions when expressed as decimal fractions are *nonterminating*. Note that the digit sequence is *cyclic* for rational, nonterminating decimal fractions. The notation of underlined digits means that the sequence of underlined digits repeats indefinitely.

Common fractions	Decimal form
1/2	0.5
1/3	0.333... = $0.\underline{3}$
1/4	0.25
1/5	0.2
1/6	0.166... = $0.1\underline{6}$
1/10	0.1
1/11	0.0909... = $0.\underline{09}$
1/25	0.04

Table 4.2.2 *Common fractions in decimal form*

Nonterminating fractions are a nuisance because it is necessary to make a decision about how many decimal places to carry in the computation. This decision can be difficult because it should be based on knowledge about accuracy of measurements used in the computation. Egyptian fractions are always terminating fractions, which is nice.

FUN QUESTION 4.2.2: Converting a common fraction into decimal form is obvious; just divide the numerator by the denominator, which in our electronic-calculator age is easy. The reverse process of converting a terminating decimal fraction into a common fraction is also obvious and easy even without an electronic calculator: 0.75 = 75/100 = 3/4. However, converting a nonterminating, cyclic decimal fraction into a common fraction is not so obvious. What is the common fraction, p/q, that produces the nonterminating decimal fraction $0.\underline{074}$?

Common fractions with denominators of the form 2^m5^n, where m and n are integers, produce terminating decimal fractions. This is because 2 and 5 are divisors (also called factors) of 10, the base. Consider the common fraction 1/2: multiply both the numerator and the denominator by 5 and obtain 5/10 = 0.5, a terminating decimal fraction. Similarly for the common fraction 1/5: multiply both the numerator and the denominator by 2 and obtain 2/10 = 0.2, a terminating decimal fraction. Thus in general $1/(2^m5^n) = 5^m2^n/10^{(m+n)}$ (see FUN QUESTION 3.2.3 for comment on power notation). Since 5^m2^n is a product of finite numbers, the numerator must be finite, and a terminating, decimal fraction results. The denominator, $10^{(m+n)}$, merely determines where to put the decimal point.

A very important advantage of decimal fractions is that it is always obvious which of two fractions is the larger. For example, 0.0123 is unambiguously larger than 0.00987. For common fractions, this differentiation is sometimes not so obvious. For example, which is larger, 13/47 or 15/51? The modern way to solve this question is simply to convert each common fraction to its decimal form on an electronic calculator. The tedious, classic way we learned as children that required finding a *common denominator* is obsolete. (Common denominator remains a useful concept in algebra and number theory, but it is obsolete in practical arithmetic.) In ancient Egypt, without electronic calculators, deciding which of a pair of unit fractions was larger, could sometimes require a tedious calculation. Which is larger, 1/70 + 1/130 or 1/62 + 1/186? (Try determining which is larger without using an electronic calculator. See the 2/*n* table, table 4.2.1, for the answer.)

Today there is not much interest in the nitty-gritty of arithmetic with obsolete Egyptian fractions, and I shall not deal with such problems. (If you are interested in calculation details, see the book by R. J. Gillings cited in the notes and references). However, the questions of *why* the method was adopted and *how* the fractions were calculated are still unresolved and of continuing interest. Let us try to solve these riddles.

A *proper fraction* has a value between 0 and 1, so when we write it as a *decimal fraction*, we only use positions to the right of the *decimal point*. Similarly, when we write a proper fraction as a *binary fraction*, we only use positions to the right of the *binary point*. Table 1.4.2 illustrated that in the binary system the position value decreases by a factor of 2 for each decrease in position. Table 4.2.3 extends this concept to positions to the right of the binary point, in other words, to negative position numbers.

Position	0	-1	-2	-3	-4	-5	-6
Position value	1	1/2	1/4	1/8	1/16	1/32	1/64
Position value, Power notation	2^0	2^{-1}	2^{-2}	2^{-3}	2^{-4}	2^{-5}	2^{-6}
Position value (decimal fraction)	1	0.5	0.25	0.125	0.0625	0.03125	0.015625

Table 4.2.3 *Unit binary fractions*

A proper common fraction with a denominator of the form 2^n can be expressed as a terminating series of unit binary fractions, which offers a clue as to the origin of Egyptian fractions. In fact, the Egyptians did use just unit binary fractions to write some common fractions: $7/8 = 0.111_2 = 1/2 + 1/4 + 1/8$; $5/8 = 0.101_2 = 1/2 + 1/8$; $7/16 = 0.0111_2 = 1/4 + 1/8 + 1/16$; and so forth. Of course, the Egyptians did not use binary notation or Hindu-Arabic symbols.

In the United States at least, binary common fractions (denominator = 2^n) expressed as a terminating series of unit binary fractions are still used nowadays, almost daily, although not consciously. Look at the markings on a ruler denominated in inches: conventionally, an inch is divided into halves, quarters, eighths, sixteenths, and sometimes even subdivided further. When we use a ruler to measure a distance, for example, of 11/16 inches, we do not bother to count eleven divisions of sixteenths. Rather, referring to figure 4.2.1, we read the ruler as a series of unit fractions, half plus eighth plus sixteenth. Thus, Egyptian use of unit fractions possibly had its origin in measurements defined by successive halving, a simple and logical method, both nowadays and thousands of years ago.

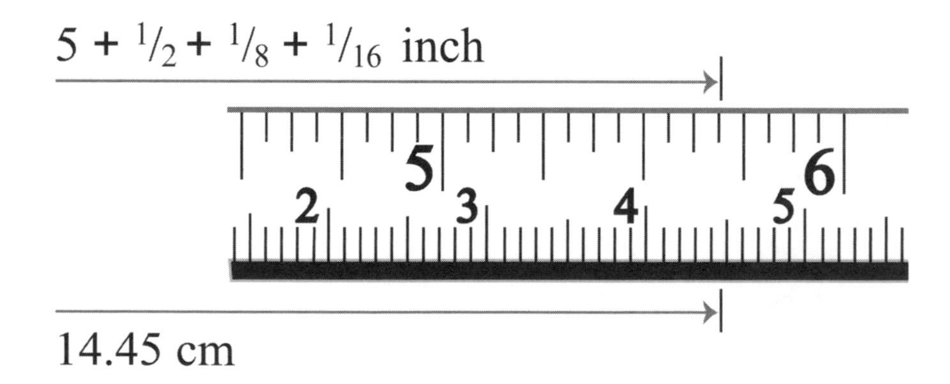

***Figure 4.2.1** Ruler markings*

For ease in reading, there is an advantage in using a length unit divided by successive halving rather than the more modern division by ten. The upper scale of the ruler in figure 4.2.1 divides inches by successive halving. The mark for each halving is always just a little shorter than the mark for the previous halving. This makes it very easy to read the fractional inch as an Egyptian fraction. The bottom scale is metric, with each centimeter divided into ten millimeters, and is somewhat less convenient to read.

Units of length are not the only ones conveniently subdivided by successive halving. Before they converted to the metric system in England, if you were very thirsty when ordering a beer in a pub you ordered a pint; if not so thirsty, you ordered a half-pint. This use of halving of units is just a peek at possibly the most extensive use of successive halving in defining units of measurement, the definitions of English units of liquid volume. This is again an ancient and natural but unconscious use of the base-2 system. As shown in table 4.2.4, the gallon is halved eleven times.

1 gallon				128 ounces	
1/2 "	**2 quarts**			64 "	
1/4 "	**1 "**			32 "	
1/8 "	1/2 "	**1 pint**		16 "	
1/16 "		**1/2 "**	**1 cup**	**8 "**	
1/32 "			**1/2 "**	**4 "**	
1/64 "			**1/4 "**	**2 "**	
1/128 "				**1 "**	**16 drams**
1/256 "					**8 "**
1/512 "					**4 "**
1/1,024 "					**2 "**
1/2,048 "					**1 "**

Table 4.2.4 *US liquid volume standards—English units*

Very large multiples or very small fractions of a unit are avoided in common usage, and new, more convenient units are defined. In this table, the units of common choice are in bold print. Which units are used is just a matter of choice: a small container of milk would be referred to as a half-pint, while a bottle of perfume of the same volume would be referred to as 8 ounces.

When we commonly think in terms of a fraction of something, we generally consider only fractions defined by successive halving. We might ask a butcher for a half or a quarter pound of meat; but the butcher would think us crazy, or at least impertinent, if we asked for 5/16 of a pound, although asking for 5 ounces is reasonable. (Not surprisingly, the English weight unit of the pound is also divided by successive halving into 16 ounces.) Even when the metric system is used, we might ask for a half or a quarter kilo; nobody would ask for a fifth kilo, although asking for 200 grams is reasonable.

The Egyptians also had volume standards defined by successive halving. A unit of volume for measuring dry grains was the *hekat*, approximately equal to a gallon. Fractions of the *hekat* were called *Horus-eye fractions* and were written with special symbols, quite unlike ordinary fractions. According to Egyptian mythology, Horus was the son of Osiris, who was treacherously slain by his brother Seth. In revenge, Horus slew his uncle, but in the fight lost his eye. The broken parts were later reassembled and his eye restored by the god Thoth. The symbols for the Horus-eye fractions are shown in figure 4.2.2. The sum of the Horus-eye fractions is only 63/64. Was the inventor of Horus-eye fractions trying to tell us that even gods could not put Humpty-Dumpty together again perfectly?

Volume units rather than length units are better examples of subdivision by successive halving because of the universal convenience of defining length units in terms of body parts. For subdivision of the inch into English units, for which there are no convenient body parts, successive halving was the natural choice. Volumes of body parts are also not easily measured, and so subdivision by successive halving was also the natural choice. As noted in section 3.1, halving of area units was also an important Egyptian operation.

However, one use of binary common fractions is surprisingly recent and has almost survived the trend to decimal fractions. Through the year 2000, stock market quotes in the United States expressed whole dollar amounts decimally, but fractional dollar values as binary common fractions. For example, in the year 2000, a quote for IBM stock was 109 3/16 and for AT&T stock it was 24 5/8. Only since the year 2001 have fractional values been expressed decimally. As I write this, the quote for IBM stock is 85.88 and for AT&T stock it is 14.25, a much more sensible method in a base-10 world. Many financial "experts" predicted chaos and opposed the conversion to decimalized stock quotes. Was this just normal fear of change or was it fear of nonterminating fractions?

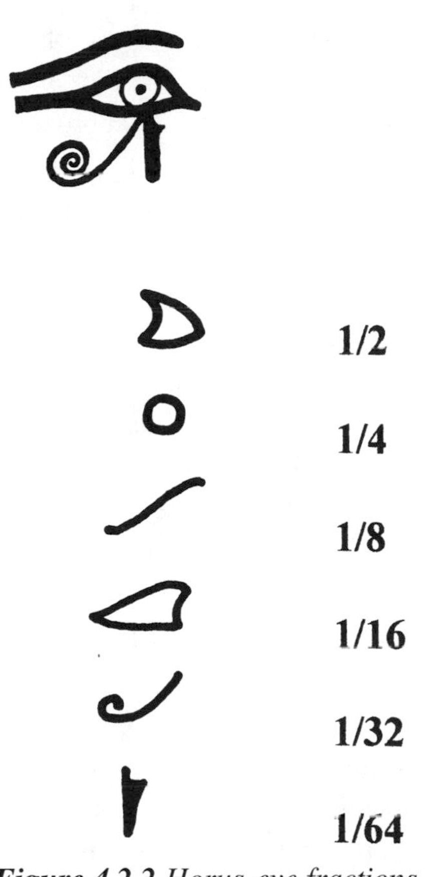

Figure 4.2.2 *Horus-eye fractions*

FUN QUESTION 4.2.3: How would you write the decimal number 5 11/16 in binary notation? Use the greedy algorithm and table 4.2.3.

FUN QUESTION 4.2.4: $5.65_{10} \times 10_{10} = ?$ $101.1011_2 \times 10_2 = ?$

When the denominator of a common fraction cannot be written as 2^n, it cannot be written exactly as a terminating series of unit binary fractions, but an approximate answer can be given to whatever precision is necessary by adding more positions. Take, for example, the fraction 19/40 = 0.475. It is a terminating decimal fraction because the denominator can be expressed as a product of divisors of ten: $40 = 2^3 \times 5$. But when expressed as a binary fraction, it is nonterminating. Truncating this nonterminating fraction as the seven-digit binary fraction 0.0111101_2 produces a decimal value of 0.477. Not a bad approximation, and probably adequate for any real problem that might have arisen in ancient Egypt, but not the Egyptian solution. They wrote 19/40 = 1/4 + 1/8 + 1/10 = 0.475, the exact value and with only three terms! The terms 1/4 + 1/8 + 1/10 no longer describe a pure binary fraction. It is a *mixed fraction*. Unfortunately, that did not bother the Egyptians because it greatly complicated their arithmetic.

In order to give some appreciation of the complications that Egyptian mixed fractions introduce, table 4.2.5 compares successive doubling of a fraction, expressed both as an Egyptian fraction and approximated by a pure binary fraction. Recall that successive doubling (see section 4.1) was the basis of Egyptian multiplication and division. Clearly, and as might have been expected, successive doubling is much easier for pure binary fractions. Doubling of pure binary fractions always produces simpler fractions. Additionally, the use of Egyptians fractions is much slower because of the constant need to resort to the 2/*n* table. If a 2/*n* table was not available, calculation was even slower. Despite awkwardness, working with Egyptian fractions is easy, and their continued use by succeeding cultures is understandable.

FUN QUESTION 4.2.5: Which binary fraction is larger, 0.0100001_2 or 0.0011111_2?

Multiplier	Egyptian fraction (conversion to unit fractions by ***Table 4.2.1***)	Pure binary fraction approximation
1x	1/5 (= 0.200)	1/8 + 1/16 + 1/128 (= 0.195)
2x	2/5 = 1/3 + 1/15	1/4 + 1/8 + 1/64
4x	2/3 + 2/15 = 2/3 + 1/10 + 1/30	1/2 + 1/4 + 1/32
8x	1 + 1/3 + 1/5 + 1/15	1 + 1/2 + 1/16

Table 4.2.5 *Successive doubling of the fraction 1/5*

To understand the origin of the use of Egyptian fractions, we must go back to the pebble-counting era where some simple problems of division of goods must have arisen. Egyptian fractions are frequently referred to as bizarre, but to appreciate how natural and intuitive they really are, let us consider how a child nowadays or in ancient Egypt, before learning much about fractions, might solve a problem of division of goods.

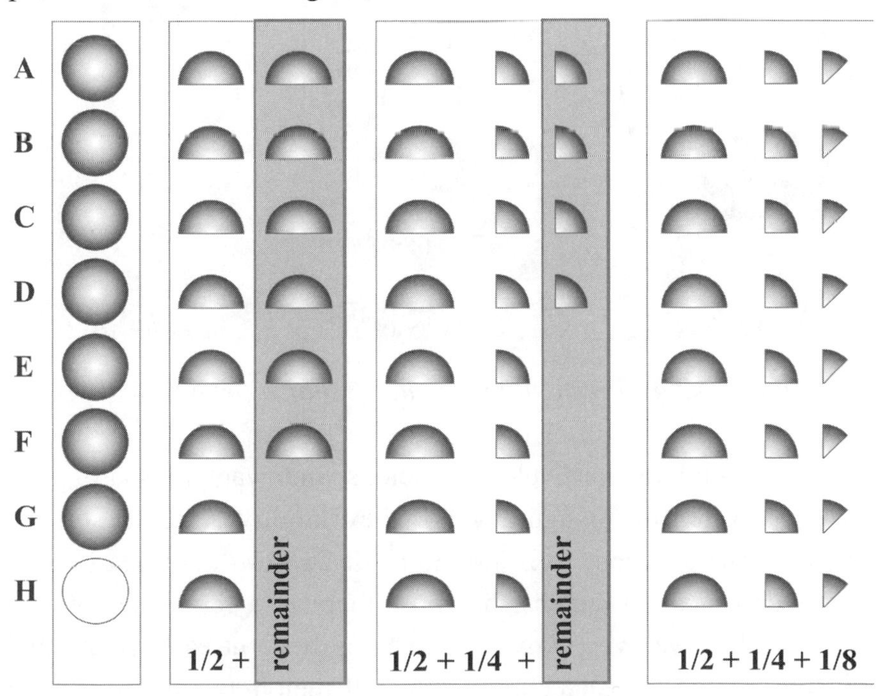

Figure 4.2.3 *Eight children equally sharing seven melons*

Eight children pick seven melons and want to share them equally. Not everybody can receive a whole melon, so, referring to figure 4.2.3, the children intuitively cut each melon in half and each child gets a half, but now there is a remainder of six half-melons. To equally share the six half-melons, each is halved and each child has an equal share of 1/2 + 1/4 melon, but now there is a remainder of four quarter-melons. Each quarter is halved and now each child has an equal share of 1/2 + 1/4 + 1/8 melon. A binary fraction gives an exact solution to this problem. Now let us consider a similar problem, but where a binary fraction cannot give an exact solution.

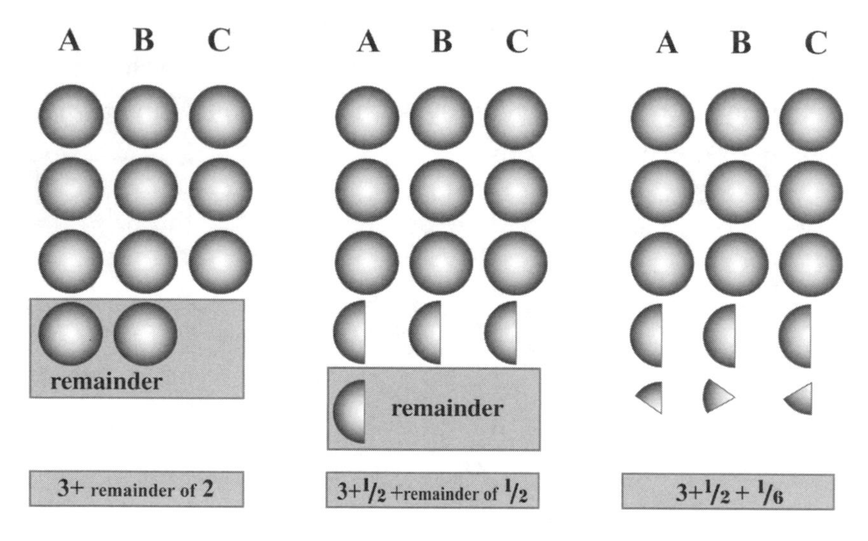

Figure 4.2.4 *Three children equally sharing eleven melons*

Three children pick eleven melons and want to share them equally. Referring to figure 4.2.4, they intuitively divide up the melons: one to A, one to B, one to C, and so forth until each child has received three melons, but now there is a remainder of two melons. To give everybody a fair share, they cut each of the two melons in half and each child gets a half-melon, but now there is a remainder of one half-melon. The process of successive division by two of the remainder can be continued until the remainder is negligible and can be simply thrown away. (This is equivalent to truncating a nonterminating fraction.) However, there is a another intuitive solution; simply cut the remaining half-melon into three equal parts, and now each child has an equal share of 3 + 1/2 + 1/6 melons.

The Egyptians presumably learned by trial and error that it was always possible to express any division and hence any fraction as a sum of a finite number of unit fractions, sometimes as pure binary fractions and sometimes as mixed fractions.

After the Egyptians began doing written arithmetic, and fractions that are more complicated were required for their more complex culture, they required a more sophisticated method of calculating unit fractions than the childlike divisions of figures 4.2.3 and 4.2.4. The Italian mathematician Fibonacci (circa 1200) provided a major clue to the *how*. In his mathematics textbook, he used the greedy algorithm to convert common fractions into a series of unit fractions. Fibonacci was of course not aware of the Rhind Papyrus, or even that the ancient Egyptians had used unit fractions; these are nineteenth-century discoveries. He was simply documenting how unit fractions, which were still in use, were actually calculated. We have seen that the Egyptians used the greedy algorithm to perform multiplication and division (see section 4.1), so its application by them to derive unit fractions is not surprising. In 1880 the eminent British mathematician James Sylvester was apparently the first to realize that the greedy algorithm could largely account for Egyptian fractions.

As an example, let's convert 19/40 into unit fractions using the greedy algorithm:

1. To find the greediest unit fraction that is less than 19/40, divide 40 by 19 and obtain 40/19 = 2 + remainder. Thus 2 < 40/19 < 3 or 1/2 > 19/40 > 1/3, so that the greediest unit fraction is 1/3.
2. Do the greedy subtraction: 19/40 – 1/3 = 17/120.

3. To find the greediest unit fraction less than 17/120, divide 120 by 17 and obtain 7 + remainder, so the greediest unit fraction less than 17/120 is 1/8.
4. Do the greedy subtraction: 17/120 – 1/8 = 1/60.
   The algorithm ends when a greedy subtraction yields a numerator of one.
5. Therefore: 19/40 = 1/3 +1/8 + 1/60

This is a simple calculation that does not require any arithmetic beyond the known capabilities of Egyptian scribes. After all, most scribes were simply bookkeepers and not necessarily mathematicians of great ability, just like bookkeepers of today. Although the greedy algorithm is not mathematically identical to the children's division-of-melons procedure, it is easy to see how the intuitive children's procedure would have led to the greedy algorithm.

This particular greedy-algorithm solution did not satisfy Egyptian scribes. The Egyptian solution, 19/40 = 1/4 + 1/8 + 1/10, is the *best Egyptian fraction* because its largest denominator is the smallest possible and therefore is the easiest to use in calculations. The rules that the Egyptians used to obtain their best unit fractions have been the subject of many mathematical investigations. However, what is most consistent with their mathematical capabilities is that starting with a greedy-algorithm solution, they simply used trial and error to modify the *greedy* result to the *best* result.

To illustrate this procedure, let's again consider the example of 19/40. The scribe could see that the greedy solution resulted in a unit fraction with a large denominator of 60 because the first two unit fractions had taken too greedy a bite. Therefore, instead of 1/3, he tried a less greedy first bite of 1/4 and obtained the best Egyptian fraction.

That the greedy algorithm can reduce any rational fraction to a finite number of unit fractions is not obvious. The Egyptians were certainly not capable of a rigorous, general proof, and so they must have simply found by experience that this was so. It is easy to prove this using simple algebra by just generalizing in algebraic notation the arithmetic of the application of the greedy algorithm to the rational fraction 19/40:

Consider a common fraction, $p/q < 1$, where p and q are integers.

1. $q/p - n +$ remainder, where n is an integer and therefore $n < q/p < n + 1$.

2. The first greedy unit fraction is therefore $1/(n + 1)$.

3. Do the greedy subtraction: $p/q - 1/(n + 1) = [p(n + 1) - q]/q(n + 1)$. The proof is to show that the numerator after the greedy subtraction is always less than the numerator before, and therefore a greedy subtraction always results in a smaller numerator. Thus, after a finite number of greedy subtractions, a numerator of unity will be reached.

4. The numerator before the greedy subtraction is p, the numerator after is $[p(n + 1) - q]$, so therefore $[p(n + 1) - q] - p = pn - q$. From **Step 1**, $pn < q$, and therefore $pn - q < 0$, QED.

Quite sophisticated mathematical analyses have been made of the unit-fraction sequences in the best Egyptian fractions. Not surprisingly, some number patterns that can be expressed as algebraic equations have emerged. The analyses of these unit-fraction sequences are interesting mathematics. However, concluding that the Egyptians used algebraic methods to derive their unit-fraction sequences, as sometimes stated, appears unjustified.

FUN QUESTION 4.2.6: Show that by using the greedy algorithm, every $2/n$ in table 4.2.1 can be expressed as a sum of two unit fractions: $2/n = 1/(m + 1) + 1/(m + 1)(2m + 1)$, where $n/2 = m + 1/2$. Note that this is not the same as saying that every $2/n$ in the table is expressed by this equation.

A drawback of mixed-base unit fractions is that there is an infinite number of different ways of writing any common fraction. This is perhaps also not obvious, and so I present a simple numerical proof for the skeptical.

1. Start with the identity: $1/n = 1/(n + 1) + 1/n(n + 1)$, where n is any integer.
2. Express any rational fraction, for example, 3/4, as an Egyptian fraction: $3/4 = 1/2 + 1/4$.
3. Use the identity of **Step 1**: $1/4 = 1/5 + 1/20$.
4. Replace the term 1/4 in **Step 2** by the expression in **Step 3**: $3/4 = 1/2 + 1/5 + 1/20$.
5. It is clearly possible to continue this process of changing and adding terms indefinitely, QED.

To help scribes find the best unit-fraction series, the Egyptians composed tables, some of which appear in the Rhind Mathematical Papyrus. However, for calculation purposes, it is not necessary to use the best Egyptian fraction. Using the best Egyptian fraction only makes calculations easier.

FUN QUESTION 4.2.7: Use the greedy algorithm to calculate the equivalent Egyptian fractions of the common fractions 2/5 and 2/9. The greedy result for 2/9 is not the best Egyptian fraction. Modify the greedy result to find the best Egyptian fraction.

FUN QUESTION 4.2.8: Use the greedy algorithm to find the equivalent binary fraction, the equivalent Egyptian fraction, and the best Egyptian fraction for the common fraction 2/27. The binary solution is nonterminating, so make the calculation precise to about 0.002.

FUN QUESTION 4.2.9: Use children's division to divide nineteen melons among forty children equally. As with any mixed fraction, there are many possible solutions for children's division, so your solution need not be the same as mine. You can easily check your result with an electronic calculator: it must equal 19/40 = 0.475.

Although the greedy algorithm—using trial-and-error changes to produce better fractions—can account for most entries in the 2/*n* table, other methods were also used, which demonstrate that the Egyptians had some nontrivial insights about properties of numbers.

Perusal of the 2/*n* table shows fewer terms were certainly preferred, but exceptions were made when even denominators could be obtained. For example, applying the greedy algorithm to 2/13, we obtain 2/13 = 1/7 + 1/91. However, changing the first unit fraction to 1/8 produces 2/13 = 1/8 + 3/104 = 1/8 + 1/52 + 1/104. Note the clever way they converted 3/104 into two even unit fractions by writing 3 = 2 + 1 and factoring 104 into 2 × 52. The Egyptians generally found even denominators preferable to fewer terms with odd denominators. However, there were limits to these preferences: no entry in the table has more than four terms and no term has a denominator greater than 980.

The Egyptians also appreciated that when *n* was not a prime number, they could exploit its factorability. Table 4.2.1 gives 2/5 = 1/3 + 1/15, the greedy solution. Then they simply calculated 2/25 = (2/5)/5 = 1/15 + 1/75, results they apparently found acceptable, even though they have odd denominators. However, 2/35 = (2/5)/7, but that is not how they calculated it. They obtained smaller and even denominators by a method that is nowadays called the *method of red auxiliaries.* When this method is used to calculate an entry in the 2/*n* table, a number in the table appears in red, which is a clue to how they calculated the entry. For the 2/35 entry, the red auxiliary is given as six. Their method was to assume a two-term solution, $2/35 = 1/a + 1/b$, so that $2 = 35/a + 35/b = (5 \times 7/a) + (5 \times 7/b) = (5 \times 7)/(5 \times 6) + (5 \times 7)/(7 \times 6) = (6 + 1)/6 + (6 - 1)/6 = 2$. Thus $a = 30$ and $b = 42$. Using combinations of the greedy algorithm, trial and error, factoring, and number smarts, the Egyptians were able to reduce any common fraction to a short sequence of unit fractions.

The Egyptians learned how to do addition and subtraction in a prehistoric, pebble-counting era. Simple multiplications and divisions were probably also mastered in this era. However, calculating complicated Egyptian fractions is a more sophisticated technique and certainly required written numbers. By about 2000 BCE, the Egyptians had an additive, written number system and knew how to use it to perform all of the essential arithmetic operations. The number system and the arithmetic were cumbersome, but in a slow-paced culture, they were adequate and used with little change for thousands of years.

4.3 EGYPTIAN ALGEBRA

Knowledge about the math of ancient Egypt has come to us by a tenuous thread of luck. First, the Rhind Papyrus, buried under sand and rubble, was found by chance. Remarkably, it had survived for about thirty-seven hundred years and was in reasonably readable condition. Then its script became decipherable through a series of fortuitous incidents. The conquest of Egypt by Alexander the Great about twenty-three hundred years ago marked the beginning of the ascendancy of the Greek culture, and the ability to read hieroglyphics eventually became a lost art and might have remained so had not Napoleon Bonaparte conquered Egypt in 1798.

Following their conquest, the French began extensive searches for and scientific study of the buried treasures of ancient

Egypt, and they started what is now called Egyptology. In 1799 one of Napoleon's officers found a stone (it measured $114 \times 72 \times 28$ cm^3) near the town of Rosetta, on which the same story about the coronation of Ptolemy V was inscribed in three different scripts, hieroglyphic, demotic, and Greek. The stone had been inscribed some two thousand years before, as the transition to the Greek language was taking place. A stone inscribed much earlier would not have contained Greek, and a stone inscribed much later would not have contained hieroglyphics.

Following the Napoleonic Wars, while the French were stealing the so-called Rosetta Stone from Egypt and shipping it to Paris, the British stole it from the French and deposited it in the British Museum, where it remains to this day. But in France in 1822, working from a copy of the inscriptions, Jean François Champollion was able to decipher the hieroglyphic text using the Greek text as the key.

Although the Rhind Papyrus was acquired in 1858, it was not until 1927 when a complete translation, *The Rhind Mathematical Papyrus* (known to cognoscenti as *RMP*), was published. This work of monumental scholarship by A. B. Chance made widely available a detailed description and analysis of at least some of the Egyptian mathematics as it was practiced around 2000 BCE. Unfortunately, the details about the site where the papyrus was found and its exact location are unknown. Such knowledge might have given some clue as to why it had survived while almost all other documents from the same era had not. Although the Rhind Papyrus appears to be only some teacher-scribe's notes, it was important enough to be copied from a document written one hundred years previously. Why the Rhind Papyrus and a few other similar documents were written, and how the peculiar mix of contents was chosen, is still a mystery.

Most of the Rhind Papyrus is devoted to examples of arithmetic solutions to practical problems. However, some problems are what we today call recreational math. The translation of *RMP* Problem 79 is:

Houses	**7**
Cats	**49**
Mice	**343**
Ears of Wheat	**2,401**
Quantity	**16,807**
Estate	**19,607**

There is no further explanation in *RMP* as to what these numbers mean. Presumably, this brief enumeration was a reminder for a teacher who told a story associated with these numbers. A clue to this story is a Mother Goose rhyme, still enjoyed by children some four thousand years later:

As I was going to St. Ives,
I met a man with seven wives.
Every wife had seven sacks.
Every sack had seven cats.
Every cat had seven kits.
Kits, cats, sacks, and wives,
How many were going to St. Ives?

The humor in this story is that, after a child starts the laborious calculation, $7 + (7 \times 7) + (7 \times 7 \times 7) + (7 \times 7 \times 7 \times 7)$, the answer is revealed as "none" because "only I was going to St. Ives."

However, especially because *RMP* Problem 79 actually gives the correct sum, it is possible that the intent of the Egyptian story was not only humor, but was also a lesson about *geometric series*.

Such a series is one in which there is a constant ratio of each term to the preceding term. Geometric series describe many phenomena: for example, compound interest produces a geometric series. Except for *RMP* Problem 79, there is no evidence that the Egyptians understood or used geometric series. But the Babylonians certainly did, so it is not unreasonable to think that the Egyptians did also.

A simple algebraic exercise will allow us to derive an equation for the sum of a geometric series. With S = series sum, a = 1st term, r = term-to-term ratio, and N = number of terms, we can write: $S = a[1 + r + r^2 + \ldots r^{(N-1)}]$, and therefore $rS = a[r + r^2 + \ldots r^{(N-1)} + r^N]$. Simply subtracting the first equation from the second, and with a little rearrangement of terms, we obtain for the sum of a geometric series:

$$S = a[r^N - 1]/(r - 1) \qquad (4.3.1)$$

The Egyptians certainly were not capable of such a derivation using algebraic notation; they must have solved *RMP* Problem 79 the hard way by successive multiplications and additions. If the derivation of this equation looks familiar, it should; it is just the algebraic generalization of the arithmetic solution to FUN QUESTION 4.2.2.

Is *RMP* Problem 79 just a humorous exercise in logic and arithmetic without any appreciation whatsoever that a geometric series is involved, or is it a tantalizing hint that the fragments of ancient Egyptian mathematics that have survived and been unearthed do not present anything like a full picture of the scope of their mathematics?

The position values for the base of any number system constitute a geometric sequence, and it is interesting to look at bases in terms of geometric series. Let us first apply equation (4.3.1) to calculate the sum of the position values of an N-digit binary number: $a = 1$, $r = 2$, and so we obtain: $S = 2^N - 1$, which is just the value deduced intuitively for the largest value of an N-digit binary number (see FUN QUESTION 1.3.1). Now let us calculate the answer to *RMP* Problem 79: $a = 7$, $r = 7$, and $N = 5$, so equation (4.3.1) yields $S = 7[7^5 - 1]/6 = 19{,}607$.

FUN QUESTION 4.3.1: Use equation (4.3.1) to calculate the largest value of a base-60, three-digit number: *X*:*X*:*X*.

FUN QUESTION 4.3.2: The position values of the base-2 fractions form a geometric series: $1/2 + 1/4 + 1/8 + \ldots 1/2^N$ of N terms. Prove that the series sum is always $1/2^N = 2^{-N}$ less than unity, and that an infinite series equals unity.

Figure 4.3.1 is a copy of the *RMP* transcription of Problem 79. At the top of the figure is a copy of the hieratic text as it appears in the original papyrus. Like cursive writing in any language, hieratic script is highly idiosyncratic and can be difficult to read unambiguously. In the lower part of the figure, Chance first transcribes the hieratic script into hieroglyphics to assure that the transcription is into combinations of hieroglyphic symbols that make sense. Then he translates the hieroglyphic transcription. Hieratics is a Semitic script, like Hebrew and Arabic, so the transcriptions of hieroglyphic numbers in the figure are read from right to left.

RPM Problem 28 is a *find-the-number-you-are-thinking problem* that is unambiguously recreational math, so *RMP* Problem 79 may also be. I have heard several variations of this problem, and they never fail to surprise the mathematically unsophisticated—which includes just about everybody. This interpretation of the problem is due to Gillings (see the notes and references) and is not Chance's interpretation in *RMP*. Here, as found in previous and subsequent chapters, the clues to the reasonable and the probably correct interpretations of ancient documents (*RMP* Problems 79 and 28) are in present-day usage. Not all mathematics may be literally timeless, but it certainly has a long memory.

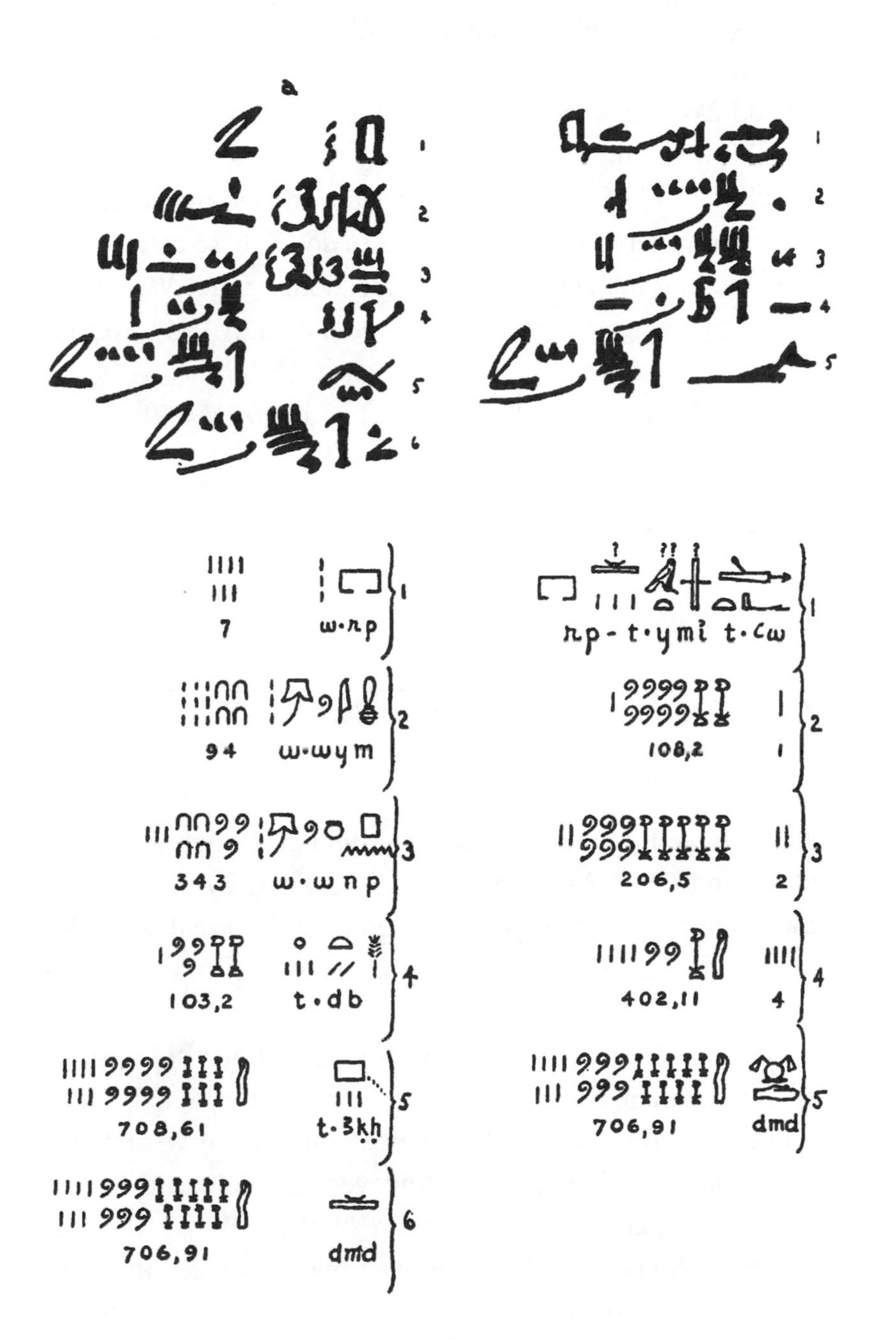

Figure 4.3.1 *Problem 79, the Rhind Mathematical Papyrus*

Someone is told: "Think of a number, N."

1. "Take 2/3 of N."

2. "Add 2/3 of N to N."

3. "Take 1/3 of the sum."

4. "Subtract 1/3 of the sum from the sum and tell me the answer, A." It takes considerable time and effort to do all these calculations.

5. On hearing the answer, A, the questioner gives N with dazzling speed by using an easy base-10 calculation: $N = A - A/10$.

To show that this really works, take the number 9 ($N = 9$).

1. $2 \times 9/3 = 6$

2. $6 + 9 = 15$

3. $15/3 = 5$

4. $15 - 5 = 10$

5. $10 - 1 = 9$

FUN QUESTION 4.3.3: Prove that for *RMP* Problem 28, $N = A - A/10$ for any N.

Although the Egyptians did not use symbolic algebra as I have (N and A), and *RMP* Problem 28 is only defined by a calculation with the same real numbers I have used here, it is really an algebraic problem with the rhetorical variables "any number" and "answer."

It is frequently stated that the ancient Egyptians did not have professional mathematicians, just scribes who invented and knew sufficient arithmetic to solve the practical administrative and engineering problems of the empire. However, as evidenced here by *RMP* Problem 28, and possibly Problem 79, there clearly also were some scribes who pondered and enjoyed mathematics without the solution of any practical problem in mind. Isn't that what mathematicians do? Thus by about 2000 BCE, the first steps were being taken from just practical arithmetic to mathematics by at least some Egyptian scribes.

* * * * *

Let us now look at what has been called the "zenith of Egyptian mathematics," an algorithm that correctly calculates the volume of the frustum of a pyramid, which is a more revealing example of Egyptian algebra. The frustum of a pyramid is just a pyramid with its top cut off, as illustrated in figure 4.3.2. It is also sometimes called a truncated pyramid. Both the Egyptians and the Babylonians were prodigious builders of pyramids. The Egyptians buried their dead under them; the Babylonians built their temples on them. However, I am not interested in the not particularly important practical uses of this calculation, but rather in the algorithm used to solve it, which shows that the Egyptians were using Babylonian algebraic technique. In that era, algebra was more an intellectual exercise than a solution to practical problems, as noted in previous consideration of *RMP* Problem 28 and of many more Babylonian examples in chapter 5. Thus, mathematical interaction between Egypt and Babylon clearly was not just business transactions with receipts and invoices, but was also an intellectual interaction.

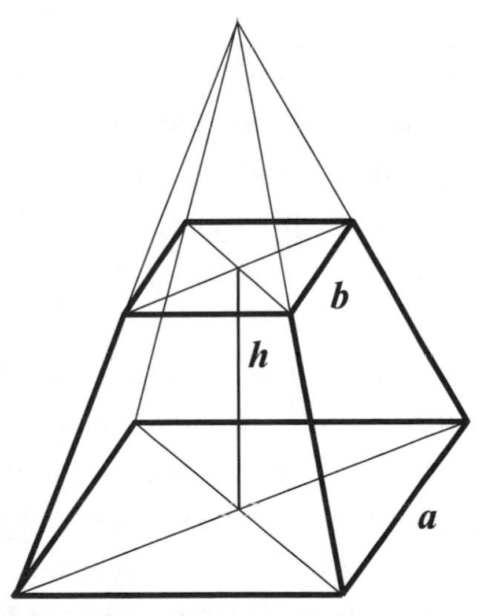

Figure 4.3.2
Frustum of a square pyramid

An algorithm that calculates the volume of a frustum is given in Problem 14 of the Moscow Mathematical Papyrus (known to cognoscenti as *MMP*), another of the very few surviving mathematical papyri from the same era as the Rhind Papyrus. Neither the Egyptians nor the Babylonians used symbols for generalized quantities (a, b, c, x, and so on) or for arithmetic operation symbols (+, –, ×, /), but rather used words. They used *rhetorical algebra* rather than our present-day *symbolic algebra*; however, I have converted the translation of the Egyptian algorithm into modern symbols.

MMP Problem 14: Calculate the volume of the frustum of a pyramid of 6 cubits for vertical height by 4 cubits on the base by 2 cubits on the top.

	Egyptian algorithm	**Algebraic generalization**
1.	$4^2 = 16$	Area of base: a^2
2.	$2^2 = 4$	Area of top: b^2
3.	$2 \times 4 = 8$	ab
4.	$16 + 8 + 4 = 28$	$a^2 + ab + b^2$
5.	$6/3 = 2$	$h/3$
6.	$2 \times 28 = 56$	$\mathbf{V(frustum) = (a^2 + ab + b^2)h/3}$

FUN QUESTION 4.3.4: Derive the Egyptian equation for the volume of the frustum of a pyramid using only high school geometry and algebra. Assume that the equation for the volume of a pyramid with a square base is $\mathbf{V = ha^2/3}$. The volume of a frustum is the difference between the volume of a pyramid and the volume of a small pyramid cut off from its top.

The equation for the volume of a pyramid is much more difficult to derive. It is questionable that the Egyptians derived it, but rather possibly obtained it empirically. It has been suggested that perhaps they compared the weight of a clay prism of height h and square base of area a^2, whose volume they knew as $V = ha^2$, with the weight of a clay pyramid of height h and square base of area a^2, and from the result concluded that $V(\text{prism}) = 3V(\text{pyramid})$.

It is possible to derive the equation for the volume of a pyramid by intuitive geometric reasoning, although nowadays the derivation is usually an elementary exercise in integral calculus. One way is by drawing the diagonals of a cube or a prism to form six pyramids of equal volume as illustrated in figure 4.3.3. Could the ancient Egyptians have been capable of such a three-dimensional visualization? While the drawing of a frustum in *MMP* Problem 14 is only a crude two-dimensional sketch, construction of interconnecting shafts and chambers within their real pyramids demonstrate three-dimensional visualization, so it is possible that they were able to derive the volume of a pyramid logically.

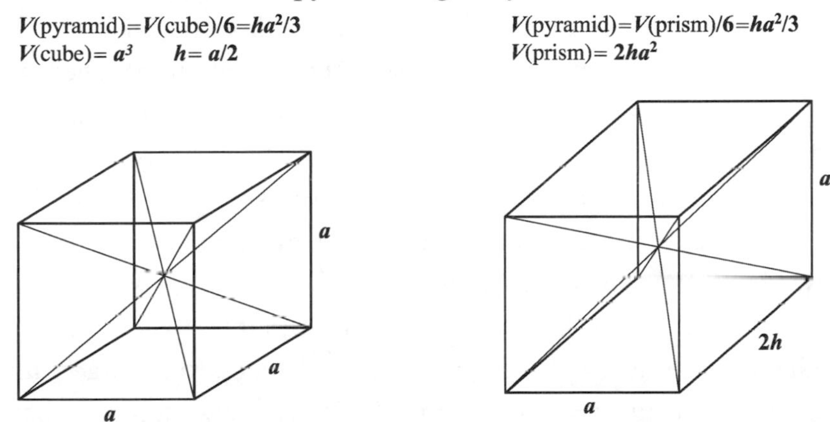

Figure 4.3.3 *A geometric derivation of the volume of a pyramid*

It is generally assumed that the Egyptians arrived at their algorithm for the volume of a frustum by purely geometric reasoning. Perhaps they did something akin to the solution of Fun Question 4.3.4, although such solutions rely on geometric concepts usually credited to the Greeks and particularly to Euclid (330–275 BCE) from more than a millennium later. This is the natural way for modern mathematicians to view the problem, but such solutions are probably more in the eyes of modern beholders than in the methods of the ancient Egyptians. An explanation more consistent with the level of Egyptian mathematics is that the algorithm was derived by a trial-and-error method, just as they obtained some of their best Egyptian fractions.

A possible Egyptian trial-and-error derivation is that a clever scribe knows that the volume of a frustum is certainly less than the volume, ha^2, of a prism with a base equal to the base of the frustum, but greater than the volume, hb^2, of a prism with a base equal to the top of the frustum. Thus, as a first guess his algorithm calculates the average, $V(\text{frustum}) = h(a^2 + b^2)/2$. Although this result has never been found in Egyptian documents, it is reasonable to assume that they knew it because it is a simple, logical guess.

A possible thought process could have been that a scribe knows that $ha^2/3$ is the correct volume of a pyramid, so he asks what he must do to his algorithm to give the correct $V(\text{pyramid})$ when $b = 0$. Instead of the divisor of 2 he tries a divisor of 3, so that his algorithm now yields $V(\text{frustum}) = h(a^2 + b^2)/3$. He now has the correct equation when $b = 0$, but otherwise it is too small. He knows it is too small because the frustum is just a prism when $b = a$, while his new algorithm yields $V(\text{frustum}) = 2ha^2/3$, only 2/3 as big as the correct value for a prism. Eventually he sees that a way to satisfy both the limits at $b = 0$ and $b = a$ is to add a step to his algorithm, yielding $V(\text{frustum}) = h(a^2 + \text{step} + b^2)/3$. The new step must have the property that it is zero when $b = 0$ and is a^2 when $b = a$. The answer is immediately obvious to him: step $= ab$. "Eureka!" he shouts, "I have just produced the zenith of Egyptian mathematics."

FUN QUESTION 4.3.5: For the frustum defined by the data of *MMP* Problem 14, what is the percent error in calculating the volume using the $h(a^2 + b^2)/2$ approximation?

The steps in the Egyptian algorithm are just the steps we would do today to evaluate numerically the algebraic expressions given in the column titled "Algebraic generalization" on page 166. Giving a numerical example was just the only way they knew to write this generalized operation. The Egyptian algorithm can be interpreted as *algebra* without *algebraic notation.*

We shall see in section 5.4 that the Babylonians similarly expressed their algebra as numerical algorithms. Thus, by around 2000 BCE there was clearly considerable diffusion of mathematics between Egypt and Babylon. Only number systems and measuring units, which had independently evolved in much earlier times before there was much interaction, were embedded in tradition and remained immune to change.

Rather than this purely algebraic, trial-and-error solution, I conjecture that the solution based on the geometric visualization of figure 4.3.4 is much more probable. In section 5.4, we shall see extensive documentary evidence from Babylon for *geometric algebra*, the composing of algorithms based on two-dimensional geometric visualizations. The algebra-without-algebraic-notation algorithm was clearly shared technique between Egypt and Babylon, and it is reasonable that geometric algebra was also shared technique.

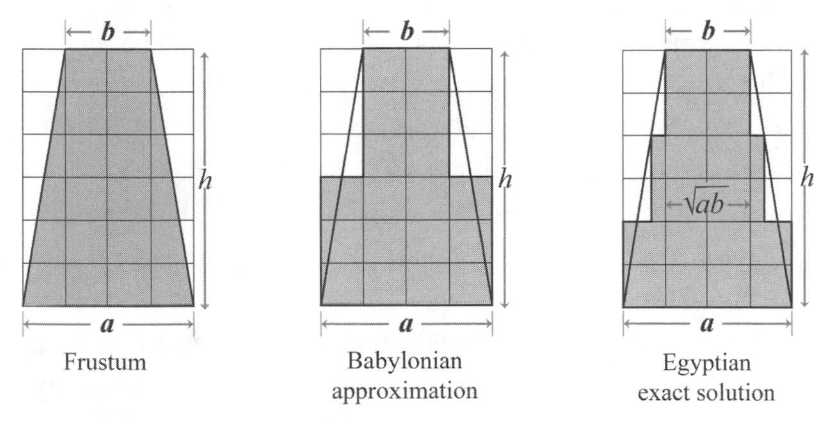

Figure 4.3.4 *Frustum volume*

The Egyptian first guess for the frustum volume was presumably the two-prism visualization. Then some clever scribe visualized that a better approximation could be obtained by approximating the frustum volume by three prisms. Clearly the base area of the middle prism must be less than a^2, but greater than b^2. The choice of ab for the base area of the middle prism certainly satisfies these requirements and is a logical, intuitive choice. Fortuitously, this choice gives the volume of the frustum exactly. The Egyptians could not even have known that this was an exact solution, but only that any deviation between calculated and observed volumes was less than the accuracy of their measurements. The middle-prism base is a square of sides $\sqrt{ab}$, which in mathematical jargon is known as the *geometric average* of a and b. Fortunately, it is not necessary to make this calculation because the Egyptians only knew how to find square roots of numbers that are integers squared (and perhaps $\sqrt{2}$; see later in this chapter).

FUN QUESTION 4.3.6: The equation $V(\text{frustum}) = h[(a + b)^2 + (a - b)^2/3]/4$ has been credited to the Babylonians. Prove that this is equivalent to the Egyptian equation.

Generally, it is considered unlikely that a truly difficult and original invention was made simultaneously and independently in two different cultures. Rather, it is usually believed that it was *invented* in only one place and obtained by the others by *diffusion*. Assuming that the Babylonians really had this equation, who first discovered the exact equation for frustum volume, the Egyptians or the Babylonians? My guess is that it was the Egyptians (see section 5.1 for further explanation of why).

What is interesting about this equation is that what was actually found on a Babylonian tablet is an algorithm that could only be algebraically generalized to $V = h[(a + b)/2]^2 + \ldots$, meaning that the remainder of the algorithm was damaged and undecipherable. Some historians have assumed that what was missing was the term required to make it equivalent to the Egyptian equation. However, B. L. van der Waerden (see notes and references) noted that the complete expression could have been: $V = h\{[(a + b)/2]^2 + [(a -$

$b)/2]^2\} = h(a^2 + b^2)/2$, just the presumed Egyptian first guess. The geometric visualization of the van der Waerden interpretation is not clear, and neither is the motivation for such a formulation. However, as we shall see in chapter 5, it was Babylonian practice to express quadratic equations in terms of $(a + b)^2$ and $(a - b)^2$ rather than just in terms of a^2 and b^2, so there is some basis to van der Waerden's conjecture. Thus, in figure 4.3.4 I have called the Egyptian first guess the Babylonian approximation. Another interpretation has been that the Babylonian equation was simply $h[(a + b)/2]^2$, differing slightly from the Egyptian first guess, but an equally logical guess. Thus, just how the Babylonians calculated the volume of a frustum is not unambiguously known.

The visualizations of figure 4.3.4 can also be looked at as precursors of the logic of the integral calculus solution for the volume of a frustum: approximating the volume as the sum of prisms, as illustrated in figure 4.3.5. In the limit as the number of prisms goes to infinity, the sum of the volume of the prisms equals the volume of the frustum. Although Newton and Leibniz only invented the calculus that allowed this visualization to be carried out in the seventeenth century, the intuitive visualization long preceded the invention of calculus, and perhaps *MMP* Problem 14 was one of the first steps in this visualization.

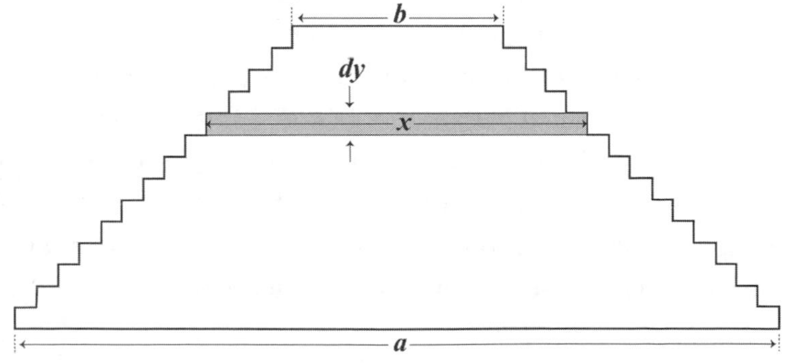

Figure 4.3.5 *The visualization of frustum volume for derivation by calculus*

FUN QUESTION 4.3.7: The calculation of frustum volume by integral calculus can be summed up as (refer to figure 4.3.5):

$$\int x^2 dy = \frac{h}{a-b}\int_b^a x^2 dx = \frac{h}{3(a-b)}\left(a^3 - b^3\right)$$

Show that this rigorous "infinite number of prisms" answer is equivalent to the Egyptian "three-prism guess." You do not need to understand calculus or the calculus notation used here to solve this problem.

Since only an extremely small fraction of ancient mathematical papyri have survived and been found, we may have a distorted appreciation of what the Egyptians knew. The "zenith of Egyptian mathematics" may appear to be abnormally outstanding only because a body of other geometry documents may have existed that have either not survived or not been recovered. The calculation of the frustum volume appears so outstanding because the easy geometric-algebra derivation given here in figure 4.3.4 has not been considered previously because it has only recently been appreciated that it even existed in Babylon (see section 5.1). However, the main reason that the Egyptian calculation of the frustum volume appears so outstanding is that modern interpreters have not realized that this result was just fortuitously exact. The watershed difference between the Egyptian result and the Greek derivation by Eudoxos (408–355 BCE) some fifteen hundred years later is that Eudoxos knew he had the exact solution.

The importance of the Moscow Mathematical Papyrus is not just for its preservation of the "zenith of Egyptian mathematics." It also alerts us to the fact that the Rhind Mathematical Papyrus, while presenting almost all that is known about ancient Egyptian mathematics, certainly does not present the complete story. What percentage of the story it presents is unknown and the general perception that it is a large percentage is unjustified.

* * * * *

There is considerable evidence of diffusion of mathematics between Egypt and Babylon, but there was Babylonian progress in geometry related to the Pythagorean theorem (see section 5.3) that is only hinted at in any Egyptian document that has been recovered. One of these hints is that the definitions of the royal cubit and the remen can be interpreted as evidence of a more profound Egyptian understanding of geometry than is generally assumed.

In section 3.1, I conjectured that the length units of the royal cubit and the remen were invented to facilitate halving and doubling of areas because $(\text{cubit}/\text{remen})^2 = (7/5)^2 \cong 2$. My guess was that this was based on the chance observation that $(7/5)^2 \cong 2$. However, the construction of figure 4.3.6 produces an exact halving or doubling of a square area. This is obvious because the square of sides c contains four triangles, while the square of sides a contains only two identical triangles. This construction is simple enough that some scribe could have chanced upon it. In the course of drawing this diagram, a scribe might also have observed that if he drew a square with sides of seven units, then using such a side as the diagonal of another square produced a square with sides of five units, to a good approximation.

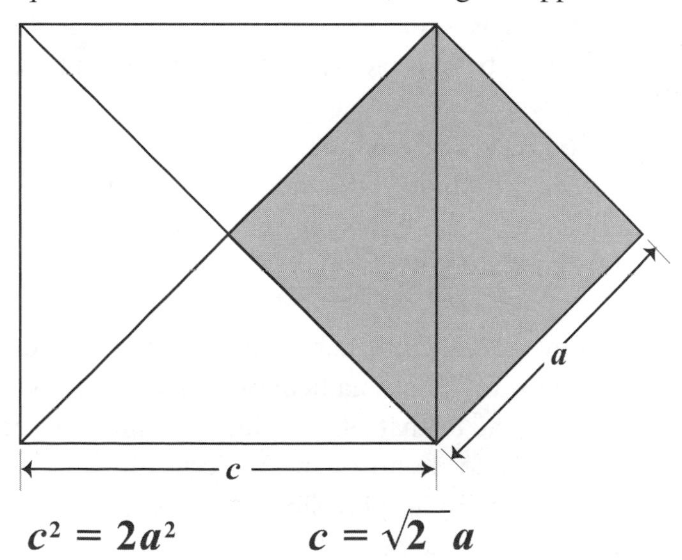

$$c^2 = 2a^2 \qquad c = \sqrt{2}\,a$$

Figure 4.3.6 *Halving and doubling of areas—exact solution*

The construction of figure 4.3.6 is not necessary to explain the invention of the royal cubit and the remen, although it could be the explanation. Its importance lies in whether it hints that the Egyptians were aware of other implications of the construction. As noted in figure 4.3.6, the construction implies that the Egyptians might have had an estimate for the square root of two: $\sqrt{2} \cong 7/5 = 1.4$, which is not a bad approximation to the correct value of 1.414. . . . There is no other evidence that the Egyptians had an estimate for the square root of two.

The most far-reaching implication of this construction is that it is actually a proof of the Pythagorean theorem, albeit only for the particular case of an isosceles right triangle. This is expressed on the figure as $2a^2 = c^2$. Essentially this same construction on a Babylonian clay tablet (YBC 7289 [YBC = Yale Babylonian Collection], see figure 5.5.1) is now widely considered as proof that the Babylonians had proved the Pythagorean theorem some one thousand years before Pythagoras.

Historians have only recently begun to digest the concept that Babylonians not only understood but possibly also proved the Pythagorean theorem. That the Egyptians were also privy to this sophisticated concept is intriguing, but it is premature to conclude that they were.

FUN QUESTION 4.3.8: What is the approximate right angle in an "Egyptian right triangle" with sides 5, 5, 7? What is the approximate right angle in an "Egyptian right triangle" with sides 7, 7, 10? You will need a little trigonometry to solve this problem.

We have now seen documentary evidence that by about four thousand years ago, Egyptian mathematics attained a level of competence about equal to that of an educated thirteen-year-old child today: an ability to do all of the basic arithmetic operations, an interest and an ability to play simple mathematical games, and an elementary understanding of algebra and geometry.

However, we have also seen hints that the pitifully small and fragmentary archeological evidence may not do justice to Egyptian mathematical achievements and that they may not have lagged behind the Babylonians as is generally assumed. Observations of the Greek historian Herodotus (484–430 BCE) support this view: "Sesostris divided the country among all Egyptians, giving each man the same amount of land in the form of a square plot. This was a source of income for him, because he ordered them to pay an annual tax. If any of a person's plot was lost to the river, he would present himself to the king's court and tell them what had happened; then the king sent inspectors to measure how much land he had lost, so that in the future the man would pay proportionally less tax. It seems to me that this was how geometry as a land surveying technique came to be discovered and then imported into Greece. But the Greeks learned about the sundial, its pointer, and the twelve divisions of the day from the Babylonians."

Unfortunately, Herodotus did not delve very deeply into mathematics, but these observations of his show that he did appreciate the debt of Greek mathematics to both prior Egyptian and prior Babylonian knowledge. It will take new archeological finds to establish that Egyptian mathematical competence has indeed been underestimated.

4.4 Pyramidiots

The credit for coining the word *pyramidiot* belongs to Leonard Cottrell, author of *The Mountains of Pharaoh* (1956). "The Great Pyramidiot" was the title of a chapter in his book about theories published in 1877 by Piazzi Smyth. One of Smyth's observations was that 1/360 of the base of the Great Pyramid of Cheops equals a twenty-millionth of earth's diameter. The base of the pyramid measures 230.6 m, so 230.6/360 = 0.64 m. As we have seen, the metric system essentially defines the circumference of earth as

40,000 km = 40×10^6 m. The diameter of earth is therefore $40 \times 10^6/\pi = 40 \times 10^6/3.1416 = 12.7 \times 10^6$ m, so that $12.7 \times 10^6/20 \times 10^6$ yields 0.64 m!

As many before me have pointed out, this calculation proves nothing. If this choice of numbers had not worked out, then one of the hundreds of other combinations of dimensions of the pyramids and the earth and other arbitrarily chosen simple integers would have yielded something similar. Nobody knows how many different combinations of numbers Smyth tried before he came up with this coincidence. Piazzi Smyth was not just an ordinary kook; he was Astronomer Royal of Scotland, which bestowed undeserved authority to his notions.

To Piazzi Smyth, the Pyramids of Giza were too colossal to have been conceived by humans, and he believed that they were "divinely inspired to serve as a compendium of weights and measures and a chronicle of man's history, past and future, which could be interpreted by mathematical calculations based on dimensions, capacities, and proportions."

The insanity of Piazza Smyth's calculation is evident when we ask what message the pyramid dimensions were revealing. The diameter of earth was unknown in the pyramid-building era. If this is what God was revealing in Smyth's equation, he clearly underestimated the ability of the Egyptians to decipher his code. They did not even understand that the earth had a diameter. By the time Smyth supposedly deciphered the code, the diameter of the earth was already long known, making the message redundant. Not very clever behavior for a presumed all-knowing God.

Piazza Smyth did not originate the concept that gods send encrypted messages, nor was he the last to claim that he was able to decipher such messages. A contemporary version of this delusion is the use of sophisticated computer scans that reveal secret messages embedded in the text of the Hebrew Bible. The primary propagator of this concept is the book by Michael Drosnin, *The Bible Code* (1998). This is another example of an insane interpretation by a man whose competence is evidenced by scientific credentials (his ability to run computer programs). An Internet search using "Bible Code" as keywords will access scholarly critiques of Drosnin's faulted probability analysis that completely invalidate his conclusions.

How scientists of apparent stature can propose such absurd concepts can be understood from the explanation given by the inmate in a mental hospital about his very clever escape attempt: "I may be crazy, but I am not stupid."

The search for encrypted messages from gods has been an ongoing occupation for millennia: from Roman priests in magnificent temples soothsaying from the shapes of entrails of disemboweled animals to Gypsies in shabby storefronts reading fortunes from patterns of dregs in teacups. The search is understandable because communication with a god is so frustratingly one-sided: we pray, but the god seldom answers. Those who believe in a god who answers prayers therefore seek his answers in encrypted signs. Those to whom God gives unambiguous vocal answers are either deified as prophets/saints or diagnosed as schizophrenics.

Certainly one source of Smyth's inspiration was simply the recent introduction of the metric system, which defined the meter in somewhat absurd cosmological terms. (In section 3.1, another zany consequence of the somewhat absurd definition of the meter was noted.) However, it is also necessary to appreciate the historical background in the nineteenth century to understand why so many people of that era, insane and sane, were preoccupied with the Great Pyramids of Egypt.

Following the recent conquest of Egypt by Napoleon, interest awakened in Europe and America about ancient Egyptian monuments and artifacts. People from all countries descended on Egypt and were busy digging up mummies, art objects, and treasures and shipping them back home to museums and private collections. The wonder evoked by these archeological finds is distilled in the poem "Ozymandias," written by Percy Bysshe Shelley in 1817:

> I met a traveler from an antique land
> Who said: Two vast and trunkless legs of stone
> Stand in the desert. . . . Near them, on the sand,
> Half sunk, a shattered visage lies, whose frown,
> And wrinkled lip, and sneer of cold command,
> Tell that its sculptor well those passions read
> Which yet survive, stamped on these lifeless things,
> The hand that mocked them, and the heart that fed;
> And on the pedestal, these words appear:
> "My name is Ozymandias, king of kings:
> Look on my works, ye Mighty, and despair!"
> Nothing beside remains. Round the decay
> Of that colossal wreck, boundless and bare
> The lone and level sands stretch far away.

What were these mysterious discoveries? How old were they? Who had made them and why? The only book available that related to these questions for most Christians was the Old Testament, which apparently was history back to the time of creation. For many it was necessary to make the new discoveries conform to cherished biblical stories.

Many people related the new finds to the biblical stories about the sojourns of Joseph and Moses in Egypt. This approach did not lead to mathematical theories, but resulted in some quaint stories. One explanation for the existence of the pyramids was that Joseph (Genesis, Old Testament; Bereshith, Hebrew Bible) had them built for grain storage during seven years of famine, in accord with his interpretation of Pharaoh's dream. This explanation evaporated, at least for the sane, when it was discovered that the pyramids were essentially solid.

In 1859, an English mathematician, John Taylor, published *The Great Pyramid: Why It Was Built, and Who Built It?* Taylor's most noted observation is that the length of the base divided by the height, for each of the three Pyramids of Giza, is equal to $\pi/2$. For example, from the dimensions of the Pyramid of Cheops: base = 230.6 m, height = 146.6 m, then 230.6/146.6 = 1.57 and $\pi/2$ = 3.14/2 = 1.57! Perhaps because no dimensions of cosmic magnitude are involved in this calculation, there have even been some competent archeologists, including the revered Flinders Petrie, the father of scientific Egyptology, who have tended to view this relationship as not just a coincidence. It was considered possible for there to have been some geometric plan involving π, although nobody has ever been able to even guess what it might have been.

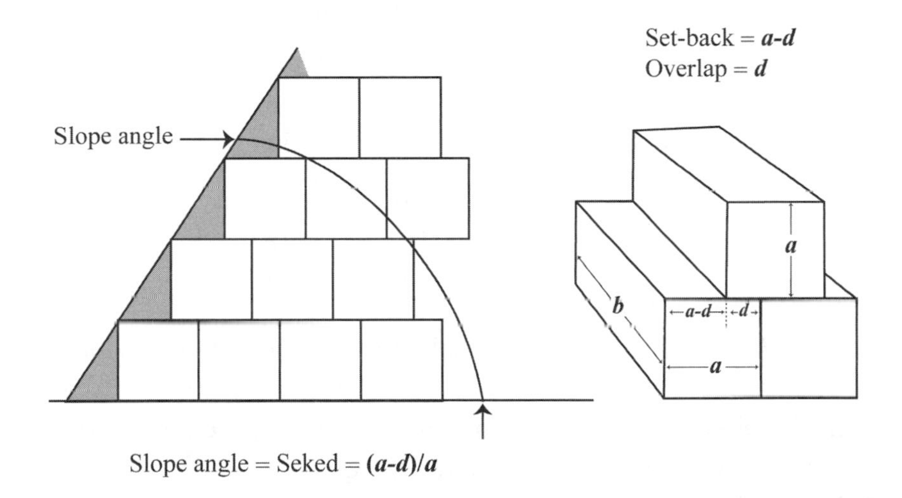

Figure 4.4.1 *Definition of slope angle and* seked *of a pyramid face*

Toward an explanation of this result as just another coincidence, let us look at how the Egyptians built the Pyramids of Giza. Figure 4.4.1 illustrates the stacking of stone blocks, prisms of length b with square ends of sides a, to construct a pyramid. For stability, they had to be stacked so that there was an overlap, d, between blocks. Without overlap, nothing holds down an outside block except its own weight. The concept of angular measure had not been invented yet, but the Egyptians determined the slope angle by exactly what had to be measured when a new layer of blocks was added: the setback, $(a - d)$, that the block should be placed from the face of the underlying block.

The slope angle was defined by the *seked*, the ratio of the setback to the height of the block: *seked* = $(a - d)/a$. (The *seked* coincides with the modern trigonometric definition of the cotangent.) Table 4.4.1 presents the slope angle and *seked* of the three Pyramids of Giza and data from four problems in the *Rhind Mathematical Pypyrus* (*RMP*).

Pyramid	**Slope-angle (degrees)**	***seked***
Cheops	51.87	0.78
Chephren	52.33	0.77
Mycerinus	50.78	0.81
RMP 56	54.23	0.72
RMP 57	53.13	0.75
RMP 58	53.13	0.75
RMP 59	53.13	0.75

Table 4.4.1 *The slope angle and* seked *of some Egyptian pyramids*

***RMP* Problem 56**: The pyramid height is 250 royal cubits, and the base is 360 royal cubits. What is the *seked*?
***RMP* Problem 57**: The *seked* of a pyramid is 5 palms 1 finger (a = 1 royal cubit), and the base is 140 royal cubits. What is the height?
***RMP* Problem 58**: The height of a pyramid is 93 2/3 royal cubits, and the base is 140 royal cubits. What is the *seked*?
***RMP* Problem 59**: The height of a pyramid is 8 royal cubits, and the base is 12 royal cubits. What is its *seked*?

FUN QUESTION 4.4.1: The *seked* of a pyramid is 2 palms 2 fingers (a = 1 royal cubit). Express the *seked* as a fraction of a royal cubit in Egyptian fraction form.

It is quite surprising to me that apparently nobody has previously perceived that the *RMP* problems suggest that there may have been a target *seked* of 0.75. A *seked* of 0.75 means that the overlap of the blocks is 25 percent. An overlap that is much smaller than this would not be stable because the outside block would not be firmly grabbed. A much larger overlap results in a slope angle that is too large, which produces a less stable pyramid or simply does not convey the intended image of stability.

The base-to-height ratio of a pyramid is twice the *seked*, so for a *seked* of 0.75, the ratio is 1.5, slightly less than $\pi/2 = 1.57$. The positioning of blocks, whose average weight was 2.5 tons (2.24 metric tons), could not have been easy. Presumably, the three Pyramids of Giza were designed to have the same slope, but we can see from the observed spread in slopes noted in table 4.4.1 that the Egyptian inability to replicate a designed slope well accounts for the discrepancy between a target *seked* of 0.75 and the observed *sekeds*.

A targeted *seked* of 0.75 is a reasonable explanation of the pyramid slopes, but whether right or wrong, there certainly are explanations based on engineering considerations capable of accounting for why the base-to-height ratio of the Great Pyramids is about 1.57 and there is no need to invoke the coincidental value of $\pi/2$. (See FUN QUESTION 5.4.2 for another zany way to read significance into pyramid slopes.)

A consequence of a targeted *seked* of 0.75 = 3/4 is that the slope is defined by a right triangle similar to a triangle with sides 3, 4, 5. Such a triangle is a Pythagorean triple; the Pythagorean theorem is satisfied with integers for all sides: $3^2 + 4^2 = 5^2$. Pythagorean triples play an important role in Babylonian and other ancient mathematics (see section 5.3); surprisingly they are missing from the Egyptian archeological record. Were *RMP* Problems 57–59 written with complete unawareness that they relate to a Pythagorean triple? That pyramid slopes can be accounted for by a Pythagorean triple appears to be just coincidental, but it may not be.

By the end of the nineteenth century, competent archeologists with their ability to read hieroglyphics had pretty much solved the mystery of the pyramids. They were simply gigantic burial monuments of pharaohs. They are also monuments to gigantic folly. Designed to prevent grave robbing, they were all pillaged within a few years after the burials. Later pharaohs were buried deep in desert hillsides without advertising, "Here I am, see if you can find my treasures," although with not much more success in foiling grave robbers. Today the exteriors of the pyramids are a checkered pattern of eroded blocks of stone. Originally the pyramids had a smooth casing of white limestone (the shaded part of figure 4.4.1), almost all of which has been stolen to use as building material. In their original condition, they must have been an awesome sight.

With the modern understanding of when, why, and who built the pyramids, it might be expected that analyses of pyramid dimensions would have ceased. Not so. Analysis continues with undiminished vigor, and surprisingly with some interesting new results. Figure 4.4.2 is from a recent study by John Legon. It shows that if the layout of the Pyramids of Giza is measured in royal cubits, the site is framed by a rectangle of dimensions 1,417.5 by 1,732, which is reasonably approximated as 1,000 $\sqrt{2}$ by $\sqrt{3}$ 1,000. Nobody has yet thought of an explanation for why these square roots are involved, but not for lack of effort. However, unlike the nineteenth-century calculations of Piazza Smyth and John Taylor, the rather good approximations of $\sqrt{2}$ and $\sqrt{3}$ do not depend on calculation of ratios, but are actual dimensions. It is improbable, but not impossible, that this is just coincidence, but let us see if what we have learned about Egyptian measurement practice can help resolve this riddle.

In section 3.1 we saw that the Egyptians tended to define land area metrically, so it is reasonable that a standard-size plot of 1,000 $\times$ 1,000 (royal cubits)^2 was initially chosen to be the site for building pyramids of the pharaoh Cheops and his progeny. However, sometime later, either before or after the Pyramid of Cheops

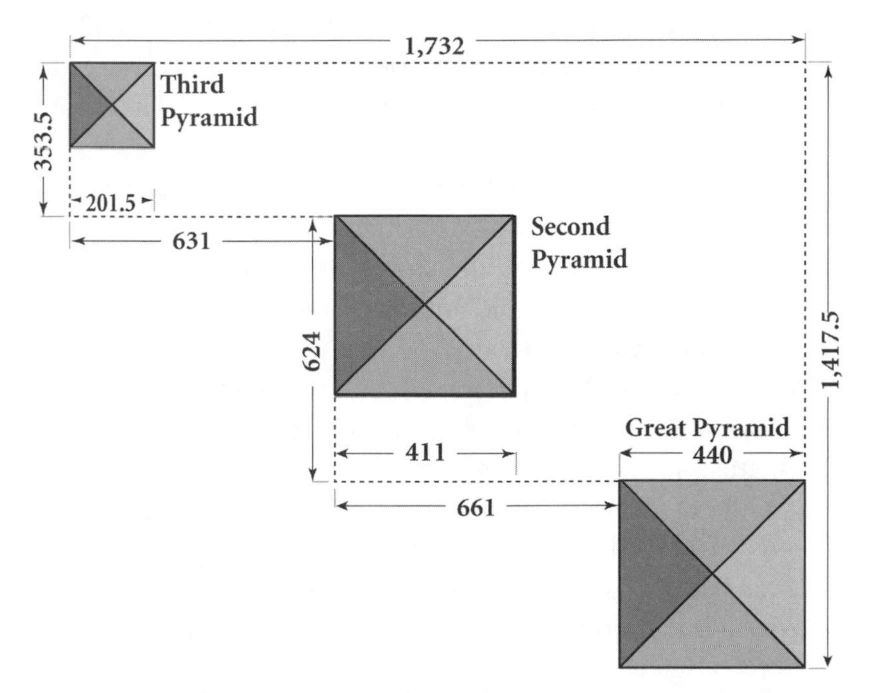

Figure 4.4.2 *Giza site plan—dimensions in royal cubits*

was built, it was decided to enlarge the plot. As noted in section 3.1, Egyptian practice was to increase areas by doubling, which we saw meant increasing linear dimensions by the ratio of the royal cubit to the remen, or by $7/5 \cong \sqrt{2}$. We thus have a rational explanation for why the burial plot would be a square of sides, $1{,}000\ \sqrt{2}$, and without requiring Egyptian knowledge of the value of the $\sqrt{2}$, which as also noted in section 3.1 was possibly not known. There also is no evidence that the value for the $\sqrt{3}$ was known.

Referring to figure 4.4.2, we see that there was clearly an attempt to align all of the pyramids along their diagonals, but it was simply not very precisely done. The placement of the Pyramid of Mycerinus is the most deviant. Whether this is due to surveying error or topographic restraints is not clear or particularly important. The net result is that the outlining plot is not a square, but a rectangle, and just coincidently one side came out approximately as 1,000 $\sqrt{3}$. That both dimensions, 1,000 $\sqrt{2}$ and 1,000 $\sqrt{3}$, are coincidental is improbable. That 1,000 $\sqrt{3}$, appears because it is just an accidental small deviation from an intended 1,000 $\sqrt{2}$ is quite reasonable.

Pyramid dimensions are precisely defined, because the Egyptians knew how to cut stone blocks precisely. However, placement of pyramids and determining the slopes of pyramids could not be done so precisely, so we should not read too much meaning into the exact values of such measurements.

The dimensions of the pyramid site have attracted some recent attention, but the alignment of the pyramids has attracted much more interest, primarily owing to the book by Robert Bauval and Adrian Gilbert, *The Orion Mystery—Unlocking the Secrets of the Pyramids* (1995). These authors present what appears to be a serious analysis of Egyptian mythology and astrological data to conclude that the alignment of the three Great Pyramids of Giza mirrors the alignment of the three prominent belt stars in the constellation of Orion. An Internet search using "Orion Mystery" as keywords will access scholarly critiques of both the book's interpretation of Egyptian mythology and the astrological data.

Arcane pyramid arithmetic does not evoke much general interest nowadays, but another pyramid theory made front-page headlines worldwide. The Maya of Central America were also builders of pyramids. Ruins of the first modest mounds to the final impressive structures clearly show indigenous development (see section 3.3). Essentially all experts agree that any resemblance to Egyptian pyramids is purely coincidental. However, a Norwegian adventurer/archeologist, Thor Heyerdahl, claimed that the Maya had learned their pyramid construction by *diffusion* from Egypt. He sought to "prove" this by showing that ancient Egyptian boat-building technology was capable of crossing the Atlantic Ocean. From Egyptian tomb paintings, he copied the ancient Egyptian boat-building technology of tying together bundles of papyrus reeds. Heyerdahl should have known that, when they sailed in the Mediterranean, the Egyptians and everybody else used boats constructed from wood, but then his voyage would not have attracted much attention.

In 1969 he constructed a boat of reeds and courageously sailed it from Africa to America. Actually his first boat, named Ra I after the Egyptian sun god, sank. Ra II made it. Was Heyerdahl really sailing with only ancient knowledge, as he claimed? Of course not; he was armed with modern knowledge of the distances involved, of weather conditions, of wind and ocean currents. The main objection to Heyerdahl's theory, and there are many, is that the Egyptian pyramid-building era was the third millennium BCE, while the impressive Mayan pyramids were built in the first millennium CE. Thor Heyerdahl's adventures are just another wacky chapter in the history of archeology, but they make a fun read.

But one kind of wacky Egyptology is peculiar to the present space-travel era. Several authors have exploited the interest and hysteria concerning alien visitors from somewhere in interstellar space to hypothesize that these aliens were godlike builders of the pyramids, among other things. An example of such a book is Erich von Däniken's *The Eyes of the Sphinx—The Newest Evidence of Extraterrestrial Contact in Ancient Egypt* (1996). These imaginative writers have found a formula for writing best sellers that superficially appear to be based on fact, but are actually just fabricated nonsense. My guess is that these authors are probably sane and laughing all the way to the bank. Incidentally, an Internet search with keywords such as "pyramid construction" will turn up numerous sites with interpretations of various pyramid dimensions, some rational and some absurd, but all testifying to a continuing fascination with the Great Pyramids of Giza that probably exceeds that for any other ancient artifact.

Caveat lector (reader beware) applies more to Egyptian archeology than to any other.

5

MATHEMATICS BY THE WATERS OF BABYLON

5.1 Babylonian Multiplication

Nowadays when we are required to calculate mentally the product 7×8 we may do this by any of several methods. If we remember this term in the multiplication table, we can give the answer as 56, rapidly and without much effort. However, if we do not do mental arithmetic often, there may be gaps in our recollection of the multiplication table, and we use various devices to fill in the gaps. Some do the mental calculation as $(10 \times 7) - (2 \times 7) = 70 - 14 = 56$ because multiplication by 10 is so simple in the decimal system. Some do the mental calculation as $7^2 + 7 = 49 + 7 = 56$ because they happen to recall squares of numbers. A child, who has not yet mastered the multiplication table, refers to a printed multiplication table that is handy in an arithmetic book. Thus, the given answer, 56 in this example, does not define whether the answer was obtained strictly from memory, from memory plus calculation, or from a printed table.

A more complicated pencil/paper multiplication, and even when written out in some detail as:

$$\begin{array}{r} 26 \\ \times \underline{43} \\ 78 \\ \underline{1{,}040} \\ 1{,}118 \end{array}$$

does not determine unambiguously how the arithmetic was done. Was 78 obtained using the standard, memorized or printed, multiplication table $3 \times 6 = 18$, $10(3 \times 2) = 60$ plus the mental addition of $18 + 60 = 78$, or was the calculation $3 \times 26 = (3 \times 25) + 3 = 78$, because $3 \times 25 = 75$ is a frequently occurring and hence memorized product (especially in the United States with its 25¢ coin)?

We base our present understanding of Babylonian multiplication on evidence that is very similar to these present-day examples. But just as with these examples, we can never know exactly how the Babylonians did multiplication from similar evidence.

Some Babylonian multiplication tables survive, and some calculations with missing intermediate steps survive. What we can add to this evidence is the understanding that a contemporary human brain functions and has capacity that is essentially identical to that of a Babylonian brain, so what was a reasonable calculation for Babylonians must also be a reasonable calculation for us. We can also expect that the Babylonians, as we do nowadays, employed various multiplication techniques. Thus, we require that any possible method be consistent with the archeological record, that it has a reasonable memorization burden, and that it is neither too tedious nor too difficult.

Archeologists have recovered more than five hundred thousand Babylonian clay tablets. The largest collection is in the British Museum, and other large collections are at Yale University, Columbia University, the University of Pennsylvania, the University of Chicago, and several other sites. Only about five hundred are of mathematical interest, and more than about two-thirds of these clay tablets are from the OB (Old Babylonian) era, 2000–1600

BCE. Developments in mathematics more or less ceased in Babylon after the end of the OB era. The mathematics of this OB era is what I shall consider here. It also includes the reign of Hammurabi, known primarily for his code, a landmark document in the evolution of moral and legal thought.

The OB period was bureaucratic, with large numbers of scribes monitoring all aspects of society. Apparently after about 1600 BCE, the bureaucracy was largely disbanded, and production of mathematical tablets diminished. Consequently, archeologists have found fewer tablets from after the OB era. This paucity of texts from later periods implies a darker age, although not quite a "dark age" because the quality of the math did not degenerate.

Babylonian base-60 counting had evolved in a prewriting era, apparently to make their number system compatible with their units of measurement (see section 3.2). In the OB era, this number system was now required to perform arithmetic that had of course not been anticipated, and, as already noted (see sections 2.1 and 4.1), arithmetic with a number system that has a large base or a large number of symbols is awkward.

Nowadays, when we do pencil/paper base-10 multiplication we use a memorizable multiplication table with only thirty-six entries (see table 4.1.1). Table 5.1.1, calculated using equation (1.3.2), presents the number of nonredundant entries in a multiplication table for each of the various bases we have encountered. Because we have ten fingers, ten was already the most widely chosen base long before ease of multiplication was a consideration. Ease of multiplication was a fortuitous bonus. In today's era of the ubiquitous electronic calculator, even in a number system with a large base, multiplication poses no problem. In Babylon it was a problem.

Base	2	8	10	16	20	60
Multiplication-table entries	0	21	36	105	171	1,711

Table 5.1.1 *Multiplication-table entries*

As a literate/numerate bureaucracy of urban scribes emerged in Babylon, its most productive move would have been to abandon base-60 and adopt base-10. Egypt, a neighbor to the west, used base-10, as did India, a neighbor to the east. OB scribes were surely aware of such use in both these countries. However, by the OB era, base-60 was embedded in the language; it was embedded in the measurement units; and it was embedded in business and government records. If the suggestion ever came up to change to base-10, it was probably resisted as vigorously as adoption of the metric system is resisted today in the United States, and for essentially the same reasons.

A possible solution to the base-60 multiplication problem that OB scribes apparently did not adopt was *Egyptian multiplication* (see section 4.1) that does not require memorizing a multiplication table. It works just as well for base-60. While no surviving artifacts suggest its use, with lack of details of intermediate multiplication steps, what evidence could have survived? Thus, some employment of Egyptian multiplication cannot be ruled out.

FUN QUESTION 5.1.1: Use Egyptian multiplication to calculate the product 13 × 173 using decimally transcribed sexagesimal and cuneiform sexagesimal notation. This is the same calculation used in section 4.1 to illustrate Egyptian multiplication.

It obviates the need for cumbersome, cuneiform multiplication tables or for much memorization. So why did the Babylonians not use it? My guess is that, if indeed they did not use it, they simply did not understand the binary number concept, which the Egyptians had fortuitously become aware of because of their general practice of halving and doubling measurement units.

The method of minimal calculation for the Babylonians would have been tables of every product from 2 × 2 to 59 × 59, which would have required the 1,711 nonredundant entries noted in table 5.1.1. However, we are sure they did not do this because no such tables have been found, but multiplication tables they did write contained a comparable number of entries. Clearly, it was not simply the bother of writing so many table entries that deterred them. My guess is that what deterred such usage was the inconvenience of lookups in a large collection of clay tablets. Neither would such a set of clay tablets have been very portable. Thus, we must look for a method with a memorizable number of entries that either eliminated the need for written tables or greatly reduced the need for them.

About 160 clay-tablet, multiplication tables—so-called *table texts*—have survived of the form: for each *principal number*, M, there is a tablet with twenty-three entries of MN for N = 1–20, 30, 40, 50. Table 5.1.2 simulates such a Babylonian $M = 10$ multiplication table. With this tablet, every product $10N$ can be obtained for $N = 1$ to 59 with at most two lookups and a simple addition. The crude writing and the quality of the tablets show that most of the surviving tablets were just student-scribe exercises and were not for use by professional scribes. Each such clay tablet is of a size that it can be held in one hand, freeing the other hand for writing.

FUN QUESTION 5.1.2: Use table 5.1.2 to multiply 10×57 in OB cuneiform.

When we consider the choices for the N entries, we note that the entries $N = 30$ and $N = 50$ are apparently redundant. Without them, for example, 10×57 can still be obtained with only two lookups in an $M = 10$ table as $(10 \times 17) + (10 \times 40)$. Two redundant entries per clay tablet is not very important, but even fewer entries per clay tablet can be obtained for N = 1–10, 20, 30, 40, 50. Now only fourteen entries per clay tablet are required, a very significant reduction in the number of entries, and still never requiring more than two lookups for any multiplication. This apparently illogical choice of entries for N supports the conjecture that these tables were probably not intended for practical use. The format of the OB multiplication

Decimal		Decimally transcribed sexagesimal		Sexagesimal cuneiform	
N	10*N*	*N*	10*N*	*N*	10*N*
1	10	1	10 (= *M*)	v	<
2	20	2	20	vv	<<
3	30	3	30	vvv	<<<
4	40	4	40	vvvv	<<<<
5	50	5	50	vvvv v	<<<< <
6	60	6	1:0	vvvv vv	v
7	70	7	1:10	vvvv vvv	v <
8	80	8	1:20	vvvv vvvv	v <<
9	90	9	1:30	vvvv vvvv v	v <<<
10	100	10	1:40	<	v <<<<
11	110	11	1:50	< v	v <<<< <
12	120	12	2:0	< vv	vv
13	130	13	2:10	<< vvv	vv <
14	140	14	2:20	< vvvv	vv <<
15	150	15	2:30	< vvvv v	vv <<<
16	160	16	2:40	< vvvv vv	vv <<<<
17	170	17	2:50	< vvvv vvv	vv <<<< <
18	180	18	3:0	< vvvv vvvv	vvv
19	190	19	3:10	< vvvv vvvv v	vvv <
20	200	20	3:20	<<	vvv <<
30	300	30	5:0	<<<	vvvv v
40	400	40	6:40	<<<<	vvvv <<<< vv
50	500	50	8:20	<<<< <	vvvv << vvvv

***Table 5.1.2** Cuneiform 10x multiplication table*

tables is better explained by treating Babylonian numbers not as simply base-60 but rather in terms of their more fundamental nature as a sequence of alternating 1-for-10 and 1-for-6 replacements (see section 3.2).

When we do a base-10, pencil/paper multiplication of an m-digit by an n-digit number using the multiplication table of table 4.1.1, whether written or memorized, we perform mn multiplications. (See, for example, the calculation of 23×43 at the beginning of this chapter where $m = n = 2$.) Treating Babylonian numbers as pure sexagesimal numbers, to multiply an m-position by an n-position number similarly requires mn multiplications. But when we note that each sexagesimal position is composed of a units digit and a tens digit, each position-by-position multiplication can be looked at as multiplication of a 2-digit by a 2-digit number. So treated, multiplying an m-position by an n-position sexagesimal number would require up to $4mn$ multiplications (less than $4mn$ when some positions do not have both units and tens digits), but far fewer multiplication-table entries are required.

A Babylonian cuneiform units-by-units multiplication table requires thirty-six entries, just as for the decimal system; a Babylonian tens-by-tens multiplication table requires fifteen entries; and a Babylonian units-by-tens multiplication table requires thirty-four entries, for a total of eighty-five entries. Tables 5.1.3, 5.1.4, and 5.1.5 illustrate the determination of the number of nonredundant entries (I am not implying that this was the format of any Babylonian tabulation; this is simply a convenient format to identify only the nonredundant entries.)

	vv	vvv	vvvv		vvvv v		vvvv vv		vvvv vvv		vvvv vvvv			vvvv vvvv v		
Symbol value	1	1	10	1	10	1	10	1	10	1	60	10	1	60	10	1
vv	vvvv	vvvv vv		vvvv vvvv	<		<	vv	<<	vvvv		<	vvvv vv		<	vvvv vvvv
vvv		vvvv vvvv v	<	vv	<	vvvv v	<	vvvv vvvv	<<	v		<<	vvvv		<<	vvvv vvv
vvvv			<	vvvv vv	<<		<<	vvvv	<<	vvvv vvvv		<<<	vv		<<<	vvvv vv
vvvv v					<<	vvvv v	<<<		<<<	vvvv v		<<<<			<<<<	vvvv v
vvvv vv							<<<	vvvv vv	<<<<	vv		<<<<	vvvv vvvv		<<<< <	vvvv
vvvv vvv									<<<<	vvvv vvvv v		<<<< <	vvvv vv	v		vvv
vvvv vvvv											v		vvvv	v	<	vv
vvvv vvvv v														v	<<	v

Table 5.1.3 *Babylonian cuneiform units × units multiplication table*

	vv	vvv		vvvv		vvvv v		vvvv vv		vvvv vvv		vvvv vvvv		vvvv vvvv v	
Symbol value	10	60	10	60	10	60	10	60	10	60	10	60	10	60	10
<	<<		<<<		<<<<		<<<< <	v		v	<	v	<<	v	<<<
<<	<<<<	v		v	<<	v	<<<<	vv		vv	<<	vv	<<<<	vvv	
<<<		v	<<<	vv		vv	<<<	vvv		vvv	<<<	vvvv		vvvv	<<<
<<<<				vv	<<<<	vvv	<<	vvvv		vvvv	<<<<	vvvv v	<<	vvvv vv	
<<<< <						vvvv	<	vvvv v		vvvv v	<<<< <	vvvv vv	<<<<	vvvv vvv	<<<

Table 5.1.4 *Babylonian cuneiform units × tens multiplication table*

	<		<<		<<<			<<<<			<<<< <		
Symbol value	60	10	60	10	600	60	10	600	60	10	600	60	10
<	v	<<<<	vvv	<<		vvvv v			vvvv vv	<<<<		vvvv vvvv	<<
<<			vvvv vv	<<<<	<			<	vvv	<<	<	vvvv vv	<<<<
<<<					<	vvvv v		<<			<<	vvvv v	
<<<<								<<	vvvv vv	<<<<	<<<	vvv	<<
<<<< <											<<<<	v	<<<<

Table 5.1.5 *Babylonian cuneiform tens × tens multiplication table*

This is a memorizable table size. Perusal of tables 5.1.3, 5.1.4, and 5.1.5 shows that many of the products are very simple, and with only two symbols in Babylonian cuneiform; many entries are in an obvious sequence. Memorizing these eighty-five entries would not have been much more of a burden than memorizing our present-day base-10 multiplication table. It would be surprising if OB scribes had not used such a method of multiplying, perhaps even with some *table text* help, because, as we saw in section 3.2, their abacus-like notation shows that they certainly understood the alternating replacement sequence character of their number system.

To illustrate the use of this method, let us again consider the calculation of 26 × 43 used at the beginning of this chapter, but this time in cuneiform notation and using tables 5.1.3–5.1.5 to evaluate the required products:

```
            <<        vvvv   = 26
                        vv
           <<<<        vvv   = 43
            <         vvvv   = 18 (3 × 6)
                      vvvv
            v                = 60 (3 × 20)
         vvvv                = 240 (40 × 6)
 <          vvv  <<          = 800 (40 × 20)
 <       vvvv <<< vvvv       = 600 + (8 × 60) + (3 × 10) + 8 = 1,118
         vvvv     vvvv
```

FUN QUESTION 5.1.3: Use tables 5.1.3–5.1.5 to calculate 52 × 37, in Babylonian cuneiform.

As we would expect from our present-day experience, using memorized multiplication tables is much faster and easier than using written tables, especially when tables are bulky clay tablets. If the archeological record shows that multiplication tables existed that enabled this calculation, we can reasonably conjecture that this was an OB multiplication procedure.

All OB multiplication tables have a format similar to that of table 5.1.2. For a set of such tables to provide the eighty-five essential entries, we require thirteen tables for M = 2–9, 10, 20, 30, 40, and 50, with each such table containing fourteen MN entries for N = 1–9, 10, 20, 30, 40, and 50. The first entry on each table, N = 1, defines M, the *principal number* of the table. Such tabulation would provide $13 \times 14 = 182$ MN entries, with many entries redundantly exhibited. We already know that the requirement for N is satisfied because existing tables have N = 1–19, 20, 30, 40, 50. The problem is just to explain why the unnecessary entries, N = 11–19, were included.

Copies of all the essential thirteen tables for M = 2–9, 10, 20, 30, 40, and 50 have survived, so there is no doubt that this method of multiplication could have been used. The problem is to explain why they bothered to write unnecessary tables for other values of M. Tables for some forty different, nonessential values of M have survived, whole or fragmented, and tables for other values of M that have not survived probably also existed.

The inclusion of the unnecessary entries on each table of N = 11 –19, adds $9 \times 13 = 117$ MN entries. If all of these also had to be memorized, memorization would have been too burdensome to be realistic. To find a product MN for an N not in the essential table, say for N = 26, the easiest and obvious way is to add essential entries $M \times 20$ and $M \times 6$. The entries N = 11–19 were therefore probably just to teach student scribes how to calculate MN entries for nonessential values of N.

Why there are tables for values of M not in the thirteen essential tables is also probably just that they were student exercises in calculating products for other values of M that were useful and were perhaps worth memorizing. Significantly, all of the multiplication tables with nonessential values of M have values of M that are *regular numbers*. Regular numbers play an important role in OB division; they will be considered in detail in sections 5.2 and 5.3. It will suffice here to note that division by regular numbers always produces terminating fractions, and the reciprocal of a regular number is also a regular number. To divide by a regular number (R), OB scribes multiplied by its reciprocal ($M = 1/R$), hence the importance of regular numbers in OB arithmetic—avoidance of nuisance, nonterminating fractions. Composing multiplication tables for the reciprocals of regular numbers not in the essential set had the dual role of teaching student scribes the importance of regular numbers in Babylonian arithmetic and of demonstrating how to multiply and divide by them. OB scribes probably memorized some of the most frequently appearing entries.

Seven is the only number in the essential set that is an irregular base-60 number. Incidentally, the Babylonian belief that seven was an unlucky number and its profound contemporary consequences (see section 3.2) is probably because division by seven produces nonterminating fractions and it is the only small base-60 number that does this. Division that results in nonterminating fractions was not just a nuisance for the Babylonians; it was generally considered impossible (see section 5.2 for more on Babylonian division).

The inclusion of nonessential tables and nonessential entries in each table has had the effect of obscuring that these tables were primarily to enable multiplication based on the fundamental nature of Babylonian cuneiform numbers as an alternating sequence of 1-for-10 and 1-for-6 replacements. The nonessential entries are much more than can be memorized and hence have led to the concept of reliance on written tables. Contributing to these misconceptions has been the lack of appreciation that most of these tables were simply student exercises and not tools of professional scribes.

Mathematicians John J. O'Connor and Edmund F. Robertson recently proposed that OB scribes invented a multiplication algorithm that required memorizing even fewer table entries. Applying the algorithm to multiply 6×7, and writing its algebraic generalization as ab, we have:

	Babylonian algorithm	Algebraic generalization
1.	$6 + 7 = 13$	$b + a$
2.	$7 - 6 = 1$	$b - a$
3.	From tables of squares: $13^2 = 169$, $1^2 = 1$	$(b + a)^2$, $(b - a)^2$
4.	$169 - 1 = 168$	$(b + a)^2 - (b - a)^2$
5.	$168/4 = 42$	$\mathbf{ab = [(b+a)^2 - (b-a)^2]/4}$ (5.1.1)

The OB way of expressing algebra was the same as the Egyptian way (see section 4.3), although, as we shall see later in this chapter, the Babylonians were most likely the originators and the Egyptians the copiers. An algorithm gave the sequence of arithmetic steps using specific numbers as an example of what should be done using any numbers, *algebra without algebraic notation.* Employing this algorithm, or equation (5.1.1), its equivalent in algebraic notation, to find all possible products of ab requires an ability to square numbers up to $(a + b)^2 = (59 + 59)^2 = 118^2 = 1{:}58^2$. The most convenient way is to have tables of squares up to $(1{:}58)^2$ with 116 entries, which does not provide easier memorization than multiplication that treats cuneiform numbers as an alternating sequence of 1-for-10 and 1-for-6 replacements, and not just as base-60 numbers.

FUN QUESTION 5.1.4: Multiply $17{:}59 \times 12{:}46$ assuming you have a table of squares up to $(1{:}58)^2 = 118^2$. To illustrate the considerable advantage in having tables of squares up to 118^2, also do the calculation assuming that you only have tables of squares up to 59^2.

One problem with the idea that OB scribes used this multiplication algorithm is that only tables of squares with fifty-eight entries from 2^2 up to 59^2 have been found, rather than up to 118^2 as required for easy use of equation (5.1.1). If such a multiplication algorithm is to be seriously considered, the Babylonians must have used a somewhat different algorithm that permitted easy calculation using tables of squares with only fifty-eight entries. This would have made it the method with minimal memorization burden, and for those scribes with a poor memory it would have provided a pocket-size multiplication calculator.

It is possible to rewrite equation (5.1.1) by simply dividing both sides of the equation by 2^2, so that it only requires a table of squares up to 59^2:

$$\boldsymbol{ab = [(a + b)/2]^2 - [(b - a)/2]^2} \qquad (5.1.1a)$$

Thus, the proposal that the Babylonians were able to perform multiplication by using tables of squares is reasonable, although probably not with proposed equation (5.1.1).

FUN QUESTION 5.1.5: There is a subtle problem with equation (5.1.1a). If $(a + b)$ is odd, $(b - a)$ must also be odd.

- Prove this. Hint: see discussion of Goldbach's conjecture in section 1.1.
- If $(b + a)$ and $(b - a)$ are odd, then $(b + a)/2$ and $(b - a)/2$ are not integers and hence their squares cannot be found in a table of squares of integers. Perhaps equation (5.1.1a) was just used when $(b - a)$ and $(a + b)$ were both even. But there are ways to enable equation (5.1.1a) to be used even when $(a + b)$ and $(b - a)$ are odd. Invent one.

There is no direct evidence that the Babylonians ever used either equation (5.1.1) or (5.1.1a), or equivalently their associated algorithms, for multiplication. That is not surprising because, just as in Egypt (see section 3.1), artifacts with the details of simple arithmetic have not been found. OB scribes and students probably did intermediate steps on soft clay that was then scraped smooth and reused, and they also used an abacus.

We know that OB scribes used multiplication tables because such tables have survived. We know that tables of squares were used because such tables have also survived, but this does not prove that they were used for multiplication. There are other uses for such tables. As can be seen in table 5.1.6, a table of squares of numbers can also be used to obtain the square root of some numbers. Examples of such use will be given in section 5.4.

$m = \sqrt{n}$	2	3	4	5	6	7	8	9	10	11	12	13	14	15
$m^2 = n$	4	9	16	25	36	49	64	81	199	121	144	169	196	225

Table 5.1.6 *Table of squares and square roots*

Noting that the Babylonians could have enjoyed the considerable advantages of a method that reduced the number of multiplication-table entries that it was necessary to memorize, and to such an extent that written tables were even portable, is not proof that they did so. What we know for sure is that equations (5.1.1), (5.1.1a), and a few others were understood because they survive as algorithms in some so-called *problem texts* (these are discussed in detail in section 5.4). But this is also not proof that these algorithms were used for multiplication.

The critical question is whether these multiplication algorithms were first invented for practical multiplication, and some problem texts are later academic generalizations of these calculations. Algebra is just generalized arithmetic (for example, the proof about the use of the greedy algorithm to calculate Egyptian fractions in section 4.2), so my conjecture is that practical multiplication algorithms based on tables of squares are the antecedent arithmetic of relevant Babylonian algebra. The very existence of relevant Babylonian algebra is thus the indirect evidence that they used such multiplication algorithms based on tables of squares.

Surprisingly, despite the large base of their number system, it was possible for the Babylonians to reduce multiplication to feasible memorization, which is how we currently do pencil/paper multiplication. Different scribes in different situations possibly used different solutions. The evidence is insufficient to determine which solutions were preferred. But we must avoid being misled by the survival of many *table texts*, with many more entries than could be memorized, into concluding that most OB multiplication required awkward lookups in a huge collection of clay tablets. Prior to our era of ubiquitous electronic calculators and computers, much numerical calculation was done with the aid of printed tables: logarithms (primarily to facilitate multiplication and division), square roots, trigonometric functions, probability functions, and so on. Even with large content per printed page and easy flipping of pages, the use of printed tables was tedious and the use of such tables did not survive for long after electronic calculators appeared. The much greater tediousness of sorting through a pile of clay tablets to perform multiplication was surely sufficient motivation to invent any of the multiplication solutions noted here—none of which were too sophisticated for the OB era. We shall see much more sophisticated OB mathematics in the next sections.

In math jargon, equations (5.1.1) and (5.1.1a) are *quadratic equations.* Babylonian algebra is not limited to quadratic equations, but I shall limit my attention to just them because they are the most interesting and misunderstood part of OB mathematics. Thus, the intuitions of mathematicians O'Connor and Robertson have probably led us to the answer of one of the outstanding questions about Babylonian mathematics: Why quadratic algebra problem texts? Because that it is what mathematicians do and apparently have been doing for some four thousand years, they generalize.

The story of the interpretation of how OB scribes derived their problem-text algorithms is a comedy of errors. Otto Neugebauer (1899–1990), a renowned history-of-mathematics scholar and translator of cuneiform mathematical texts, concluded that OB scribes derived their problem-text algorithms *algebraically*. Neugebauer gave an algebraic generalization of each step in his translation of problem-text algorithms, just as I have done recently here for a multiplication algorithm. Indeed, this *generates* algebraic equations, but it was misinterpreted as *deriving* algebraic equations. Neugebauer's stature was such that, from the time he first succeeded in deciphering a cuneiform, quadratic algebra problem text in 1927, just about every historian of Babylonian mathematics simply accepted his interpretation as unchallengable gospel.

Scholars should have realized that an algorithm is just the way we numerically evaluate an equation and is not a derivation. Nowadays when we do a numerical, pencil/paper evaluation of an equation, we usually do not consciously realize that we are using an algorithm. Only when we write a computer program to evaluate an algebraic equation numerically must we consciously pay attention to the sequence of operations and realize we are writing an algorithm. Some historians have naïvely remarked how modern and computerlike the Babylonian algorithms appear, implying some undeserved mathematical precocity.

When Neugebauer first deciphered a quadratic algebra problem text in 1927, electronic computers had not been invented yet. It is only since the 1960s, when digital computer use became widespread, that the concept of an algorithm as a numerical evaluation became widely appreciated. Working in a precomputer era and preoccupied with the daunting task of deciphering cuneiform, we can excuse Neugebauer his mistake and be thankful for his remarkable scholarship, of which we shall see more in section 5.3.

Some electronic calculators have an ability to remember a keystroke sequence, and once you numerically evaluate an equation for one set of input numbers, you can evaluate it for other sets of input numbers just by entering them in the same sequence. This is essentially how the Babylonians and Egyptians used their algorithms without algebraic notation.

Since the Babylonians did not have algebraic notation, how did they derive the "equations" their algorithms were evaluating? There is really only one possible answer; their "equations" were geometric diagrams. Such visualization of a geometric diagram is referred to as *geometric algebra*. Thus, we can read equation (5.1.1) as the Babylonians no doubt did: the area of four rectangles, each of area ab, is equal to the difference between the area of a square of sides $b + a$ and the area of a square of sides $b - a$. Figure 5.1.1 exhibits this relationship.

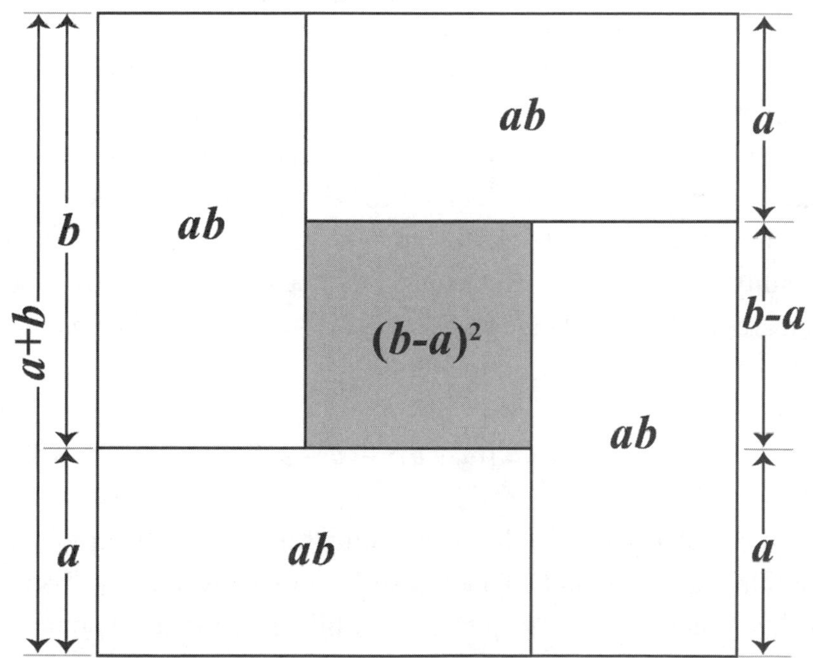

Figure 5.1.1 *Construction of a square of sides* a + b *from four rectangles of sides* a *and* b

From this figure, we can see that the area of the outer square is equal to the area of the central square plus the area of the four rectangles. Writing this in algebraic notation, we have $(a + b)^2 = (b - a)^2 + 4ab$, which with a little rearrangement yields equation (5.1.1). The diagram that relates to equation (5.1.1a), which as we have seen is a more practical equation, is simply figure 5.1.1 with all linear dimensions halved.

The diagram of figure 5.1.1 represents a very practical construction technique used by bricklayers and stonemasons, today and for thousands of years, for making a square column with a hollow center for the purpose of saving material or for building a chimney. Possibly some clever OB architect-scribe realized that such a diagram could be exploited to generate a multiplication algorithm because, as noted in section 1.1, he naturally and intuitively visualized multiplication as the area of a rectangle.

Another multiplication algorithm proposed by O'Conner and Robertson can be derived from the diagram of figure 5.1.2 The upper figure is just a small variation on figure 5.1.1, with the square hole in the center now changed to a^2. By visualizing sliding the central square down to the lower right-hand corner, it is now more readily seen that the outside square can be divided into two smaller squares and two rectangles such that $(a + b)^2 = a^2 + b^2 + 2ab$, which can be simply rearranged to yield:

$$\boldsymbol{ab = [(a + b)^2 - a^2 - b^2]/2} \tag{5.1.2}$$

For calculating products, this equation suffers from the same problem as equation (5.1.1); it also is more conveniently evaluated using tables of squares up to 118^2, while only tables of squares up to 59^2 existed. This problem can be solved by using the diagram of figure 5.1.3, which is again but a small variation on figure 5.1.2; the outside square is now changed to b^2. Again, visualizing sliding the

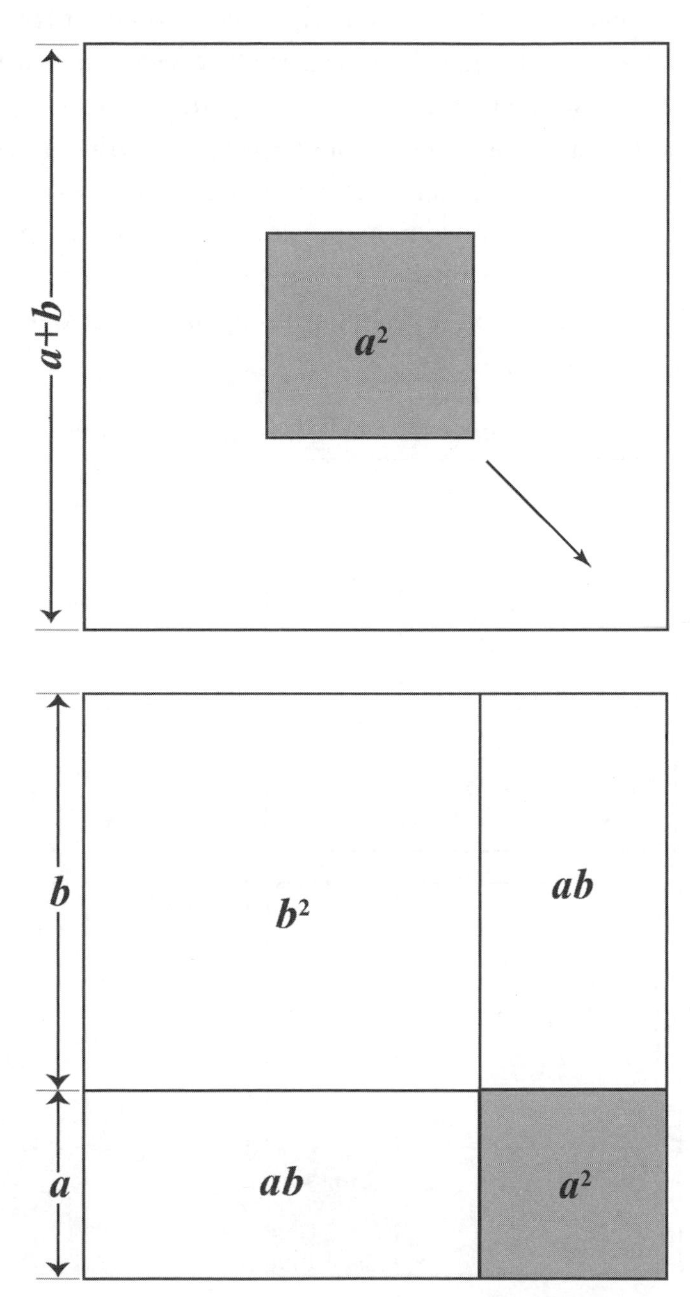

Figure 5.1.2 *Division of a square of sides* a + b *into areas* ab, a^2, *and* b^2

central square down to the lower right-hand corner, it is now more readily seen that the outside square can be divided into two smaller squares and two rectangles: $b^2 = a^2 + (b - a)^2 + 2a(b - a)$. The diagram also shows that $a(b - a)$ can be written as $ab - a^2$. Although $a(b - a) = ab - a^2$ is trivial algebra with algebraic notation, without such notation the Babylonians presumably obtained the result by the visualization illustrated. Combining these two results and with a little rearrangement, we obtain an equation that only requires tables of squares up to 59^2:

$$ab = [a^2 + b^2 - (b - a)^2]/2 \quad (5.1.3)$$

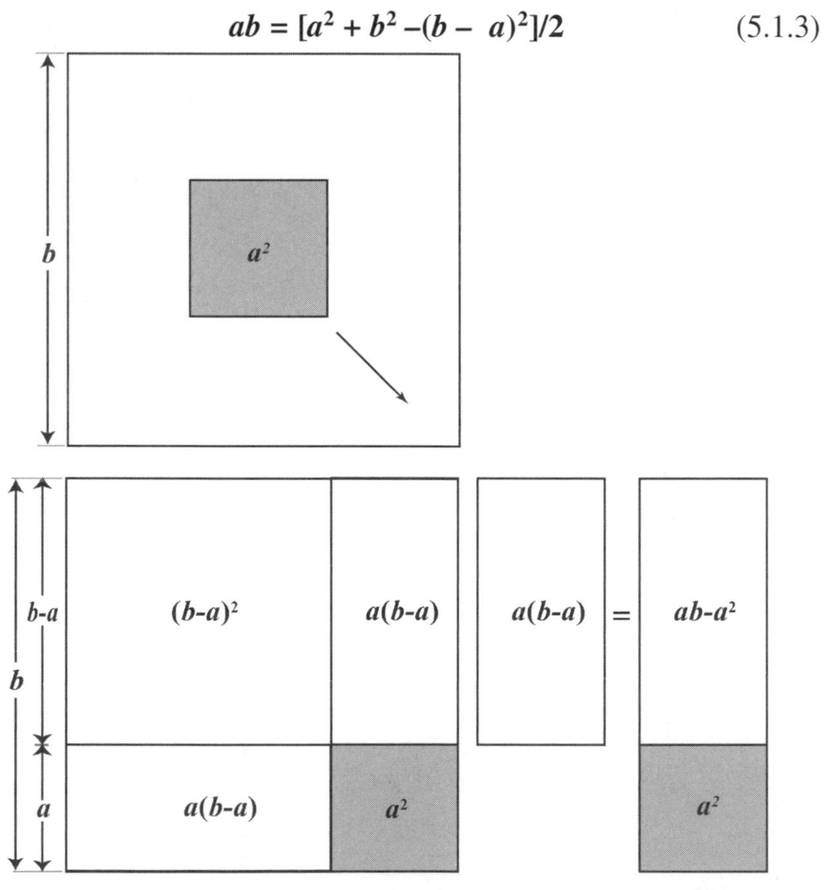

Figure 5.1.3 *Division of a square of sides* b *into areas* a(b – a), a², *and* (b – a)²

This equation is preferable to equation (5.1.1a) when $(b - a)$ is odd because no special treatment is required for this case. However, when $(b - a)$ is even, equation (5.1.1a) gives a quicker solution because it requires only two lookups in a table of squares rather than the three required by equation (5.1.3).

FUN QUESTION 5.1.6: Show that $V(\text{frustum}) = h(a^2 + ab + b^2)/3$, the Egyptian equation for the volume of a frustum, can be converted to $V(\text{frustum}) = h[(a + b)^2 + (a - b)^2/3]/4$, the Babylonian equation, by using equations (5.1.1) and (5.1.2).

As shown in section 4.3, the derivation of the Egyptian equation for the volume of a frustum is easy to understand. On the other hand, it is difficult to see how the Babylonian equation could have been derived except by starting with the Egyptian equation.

FUN QUESTION 5.1.7: Derive equation (5.1.1) by combining the geometric algebra of figures 5.1.2 and 5.1.3 rather than from figure 5.1.1. We can be reasonably sure that the Babylonians used geometric algebra, but we cannot be certain just what geometry was visualized.

It is remarkable that scholars did not realize much earlier that the Babylonians derived their quadratic equations from two-dimensional visualizations like figure 5.1.1, despite Neugebauer's contagious blind spot. Such visualizations have long been known as *Pythagorean geometric algebra.* We now appreciate that Pythagoras probably learned these constructions in Babylon, where OB scribes had invented them more than one thousand years earlier.

In section 4.3, we saw that a two-dimensional visualization was most probably the way the Egyptians derived the algorithm, or equivalently the equation, for the volume of the frustum of a pyramid, "the zenith of Egyptian mathematics." Like the Babylonian multiplication equations just considered here, the Egyptian equation is also quadratic and its geometric-algebra derivation is not inferior to any Babylonian derivation. The wider attention to quadratic geometric algebra in Babylon was presumably because of their unique problem of multiplication with a large base. This motivated the Babylonians to develop an extensive set of problem texts to teach quadratic geometric algebra to student scribes. The large body of Babylonian quadratic algebra problem texts that has no counterpart in Egypt has created a questionable impression that the Babylonian mathematics was far superior. There was simply no reason for such extensive attention to this technique in Egypt; multiplication was not a problem for the Egyptians.

While the generalization of practical multiplication algorithms probably led to algebraic problem texts, the generalized geometric visualization learned from the problem texts led to some remarkable new arithmetic, an arithmetic/algebra feedback cycle. In section 5.3, we shall see how Babylonians exploited the visualization of figure 5.1.1 to calculate Pythagorean triples, and in section 5.5, we shall explore how they exploited the visualization of figure 5.1.2 to calculate square roots. These two calculations are "the zeniths of Babylonian mathematics." Possibly, neither of these second-millennium BCE calculations would have been invented had the Babylonians not made an unfortunate and unintended choice of a number system with a large base around the fourth millennium BCE. Adversity (but not too much) is the mother of invention.

5.2 Babylonian Fractions

One and probably the only advantage of base-60 is that it has many divisors (see section 3.2), thereby allowing simple conversion of some divisions into easier multiplications. However, at best, OB division cannot be simpler than OB multiplication, which is more awkward than multiplication with base-10 because the large base requires more memorization and perhaps some use of written tables.

Just as in base-10, division in base-60 somctimes can produce nuisance, nonterminating fractions. Terminating fractions are comfortable to work with because they are always convertible into integers, and are thus essentially equivalent to integers. For example, suppose a measurement yields 3.5 m, a terminating, base-10 *noninteger*. The measurement in centimeters converts the answer to 350 cm, an *integer*. Without the concepts of a sexagesimal point and a zero symbol, there is not even any way of distinguishing between an integer and a terminating fraction in OB cuneiform, except from the context of the text.

Egyptian scribes avoided nonterminating fractions with their mixed-base unit fractions (see section 4.2). OB scribes simply avoided any division that produced nonterminating fractions. OB scribes also constructed tables of *reciprocals*, so that multiplication by reciprocals could replace division. The usual criterion for an acceptable division was that the reciprocal was either an integer or equivalently a terminating fraction. Incidentally, the modern solution to nonterminating fractions is the electronic calculator. It effortlessly calculates nonterminating fractions to far greater precision than is normally required, and we can ignore having to decide how many decimal places to keep—at least until the end of the calculation.

In order to understand OB reciprocal tables, we must first define some simple, basic properties of numbers. The number 60 is a *composite* number, which means it can be divided into a pair of divisors (in math jargon, it can be *factored*) as, for example, 60 = 6 × 10. As shown in the accompanying diagram, a number is factored into a pair of divisors (also called *factors*). Then each of these factors is factored into a pair of divisors, and so on until numbers that cannot be factored further are obtained. Numbers that cannot be factored further are *prime* numbers.

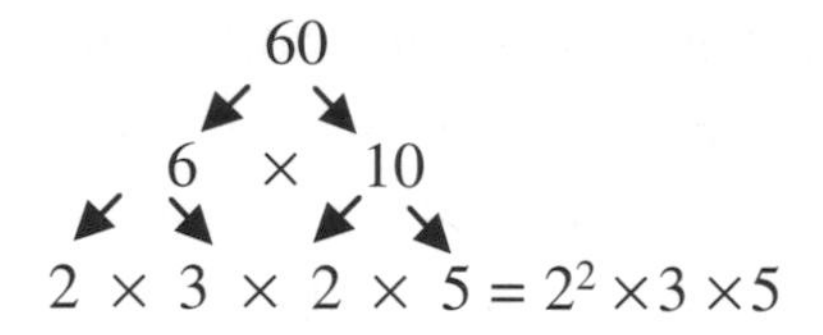

We can generalize from this factorization diagram: every number is either a composite number that can be written as a product of prime numbers or is a prime number. From the diagram it is clear that the *prime divisors* of 60 are 2, 3, and 5. In math jargon, a number that can be written as $2^i3^j5^k$, where i, j, k are integers, is a regular base-60 number. Similarly, 2^i5^k is a regular base-10 number, and 2^i is a *regular* base-2 number.

If a common fraction, p/q (p and q are integers), has a denominator that is a base-60 regular number, $q = 2^i3^j5^k$, and we multiply numerator and denominator by $30^i20^j12^k$ (since $2 \times 30 = 3 \times 20 = 5 \times 12 = 60$), we obtain:

$$p/q = p(30^i20^j12^k)/60^{(i+j+k)} \qquad (5.2.1)$$

The new denominator determines where to place the sexagesimal point. The new numerator is now the product of integers, a finite number, and hence p/q is a terminating sexagesimal fraction when the denominator, q, is a regular number. We can now tersely restate in math jargon the OB way to avoid nonterminating fractions: divide by regular numbers only.

Table 5.2.1 presents all of the reciprocals of regular base-60 numbers found in surviving *table texts*. The columns headed ***i, j, k***, and **$2^i3^j5^k$ base-10** are just to make the numbers easier to understand, and were certainly not included in OB tables. The columns headed with **base-60** are in decimally transcribed cuneiform. Many reciprocal tables have been unearthed with regular number entries from 2 to 81, but a tablet denoted as Plimpton 322 (to be discussed in section 5.3) exhibits regular numbers up to 125 and so I have included entries up to 125 in my tabulation in table 5.2.1.

Dividing only by regular numbers does not necessarily make division noncumbersome, just less cumbersome. Take, for example, division by 27, a base-60 regular number; whatever number it divides, the result is either an integer or a terminating fraction. Nonetheless, table 5.2.1 shows that replacing division by multiplication by a reciprocal still requires multiplication by a three-position sexagesimal number.

i	j	k	$2^i3^j5^k$ base-10	$2^i3^j5^k$ base-60	reciprocal base-60	i	j	k	$2^i3^j5^k$ base-10	$2^i3^j5^k$ base-60	reciprocal base-60
0	0	0	1	1	1	5	0	0	32	32	0.1:52:30
1	0	0	2	2	0.30	2	2	0	36	36	0.1:40
0	1	0	3	3	0.20	3	0	1	40	40	0.1:30
2	0	0	4	4	0.15	0	2	1	45	45	0.1:20
0	0	1	5	5	0.12	4	1	0	48	48	0.1:15
1	1	0	6	6	0.10	1	0	2	50	50	0.1:12
3	0	0	8	8	0.7:30	1	3	0	54	54	0.1:6:40
0	2	0	9	9	0.6:40	6	0	0	64	1:4	0.0:56:15
1	0	1	10	10	0.6	3	2	0	72	1:12	0.0:50
2	1	0	12	12	0.5	0	1	2	75	1:15	0.0:48
0	1	1	15	15	0.4	4	0	1	80	1:20	0.0:45
4	0	0	16	16	0.3:45	0	4	0	81	1:21	0.0:44:26:40
1	2	0	18	18	0.3:20	1	2	1	90	1:30	0.0:40
2	0	1	20	20	0.3	5	1	0	96	1:36	0.0:37:30
3	1	0	24	24	0.2:30	2	0	2	100	1:40	0.0:36
0	0	2	25	25	0.2:24	2	3	0	108	1:48	0.0:33:20
0	3	0	27	27	0.2:13:20	3	1	1	120	2:0	0.0:30
1	1	1	30	30	0.2	0	0	3	125	2:5	0.0:28:48

Table 5.2.1 *Regular base-60 numbers and their reciprocals*

Does the existence of such reciprocal tables show that OB scribes understood the concepts of prime factors and regular numbers? Not necessarily. I calculated the values in table 5.2.1, which duplicates the OB results, by using equation (5.2.1) (with $p = 1$) and used the concepts of prime factors and regular numbers. This is an easy way; calculate all possible regular numbers by choosing all possible combinations of the indices (i, j, k) and then calculate reciprocals only for them. Another way is simply to divide 1 by every number from 2 to 125, and retain for the table only those divisions that produce terminating fractions. Such division in Babylon used the intuitive greedy algorithm, just as we do nowadays in pencil/paper long division. (See FUN QUESTION 1.3.3 and the discussion that follows it). It is certainly a very tedious calculation with Babylonian sexagesimal numbers, but it is only necessary to do it once to compose a table of reciprocals.

My guess is that tedious division was the way calculation of reciprocal tables began, but then OB scribes became aware of a much easier method of calculating regular numbers and their reciprocals, which is the method that probably accounts for most of the entries in their reciprocal tables. The clue to the use of this method is a problem text, a clay tablet denoted as VAT 6505 (VAT = Vorderasiasche Abteilung Tontafeln, Berlin Museum).

VAT 6505 ostensibly presents a method for finding the reciprocals of large regular numbers that are not in existing reciprocal tables. A. Sachs, who deciphered VAT 6505, called the method simply "The Technique." The crux of "The Technique" is the insightful OB recognition that every large regular number, R, is the product of two smaller regular numbers, $R = R_1R_2$, and that the reciprocal of R can then be calculated by simple multiplication as $1/R = (1/R_1)(1/R_2)$, when R_1 and R_2 and their reciprocals are found in existing reciprocal tables. Therefore, a short table of small regular numbers and their reciprocals, perhaps calculated by tedious division, could be readily extended without limit.

FUN QUESTION 5.2.1: Prove that if R_1 and R_2 are regular sexagesimal numbers, then $R = R_1R_2$ is also a regular sexagesimal number.

By successive factoring of any large regular number, taking the number 60 as an example, we obtain (as we did previously when we used 60 as an example of a composite number):

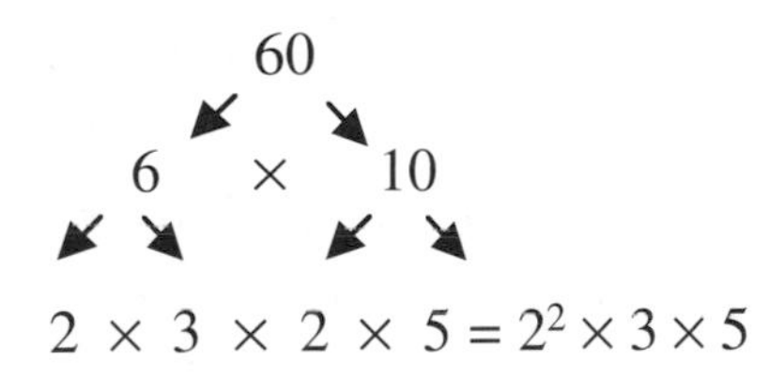

We can generalize this result to conclude that any large regular number can be written as a product of the prime factors of the base. We do not know if the Babylonians carried the insight of VAT 6505 this far so that they could essentially have composed their tables of regular numbers using essentially the same method I used. If not actually recognizing the concept of prime numbers, they certainly came very close. Greek mathematicians are generally credited with being the first to understand and appreciate the importance of prime numbers more than fifteen hundred years later (see section 6.3 for more on prime numbers).

The set of regular numbers is far smaller than the set of all numbers, and so generation of a large regular number by any practical calculation or measurement would be rare. For example, in the base-10 number system, although 60 percent of the numbers up to 10 are regular, it gives a misleading impression of occurrence. Only about 30 percent of the numbers up to 100 and only about 10 percent of the numbers up to 500 are regular. "The Technique" has been interpreted to be a practical method for factoring a large regular number that was not found in existing reciprocal tables. But since any large regular number not in existing reciprocal tables had most probably been produced by calculation as a product of two smaller regular numbers in existing reciprocal tables, "The Technique" had no practical use. Contrary to Sachs's interpretation, VAT

6505 was really just a contrived exercise in using reciprocal tables and concepts of factoring and reciprocals of regular numbers for student scribes. In section 5.4, we shall see other examples of problem texts as contrived problems for students.

Even though I have dismissed practical application of "The Technique," it is nevertheless an insightful exercise and worth understanding. One of the calculations in VAT 6505 is to find the reciprocal of the regular number < vvvvv <<<< = <<<<16:40 = $1{,}000_{10} = R = R_1R_2$. The problem was presented to students as if R_1 and R_2 were unknowns, but of course they were known and used to calculate R as R_1R_2 by the teacher-scribe who contrived the exercise. The algorithm that solves this problem is:

1. Let $R_1 = 6{:}40 = 400_{10}$ because it appears in reciprocal tables. From table 5.2.1, $1/(0.6{:}40) = 9$, which is equivalent to $1/(6{:}40) = 0.0{:}9 = 1/\mathrm{R}_1$.
2. Now the other factor can be found as $R_2 = (1/R_1)R = (0.0{:}9)(16{:}40) = 2.30 = 2.5_{10}$. However, rather than calculating R_2 this way, easier multiplication is obtained by a clever reformulation: $R_2 = (1/R_1)[(R - R_1) + R_1] = (1/R_1)(R - R_1) + 1 = (0.0{:}9)(10{:}0) + 1 = 2.30$.
3. From table 5.2.1, $1/(0.2{:}30) = 24$, which is equivalent to $1/(2.30) = 0.24 = 1/R_2$. Thus $1/R = (1/R_1)(1/R_2) = (0.0{:}9)(0.24) = 0.0{:}3{:}36 = 0.001_{10}$.

FUN QUESTION 5.2.2: Show that $125_{10} = 2{:}5_{60}$. Show that $1/125_{10} = 0.0{:}28{:}48_{60}$ without using the 125_{10} entry in table 5.2.1.

FUN QUESTION 5.2.3: How would a lazy OB scribe have divided by 49?

Although the ability to factor large regular numbers had dubious practical value, the reverse process of the ability to generate large regular numbers that were not in standard tables was a useful technique. Suppose it was necessary to divide by 251. Generally, a Babylonian would have been satisfied with the approximate answer of division by the close regular number, 250, which was not in standard regular number tables but was easily calculated as a product of two table entries.

5.3 Plimpton 322—The Enigma

Now that we know how OB scribes did multiplication and division, we are prepared to take a critical look at Plimpton 322, probably the most discussed, the most important, and the most enigmatic OB clay tablet ever unearthed. Just about every mathematician and historian who has tried to decipher or interpret Babylonian mathematics has been tempted to explain Plimpton 322. Like Oscar Wilde, I can resist everything except temptation and so I, too, shall attempt to solve the enigma of Plimpton 322.

Figure 5.3.1 is a drawing of Plimpton 322. Like most Babylonian mathematical tablets, it is small enough to hold in the palm of a scribe's hand. It is easy to discern that it is a table of columns of numbers. Most obvious is Column IV, which is just a listing of row numbers 1, 2, . . . 15, and is of no further interest. The left side of the tablet is broken off somewhere in Column I. This has created a small enigma about what exactly were the values in Column I, and a big enigma that has generated reams of speculations about what the missing part might have contained.

Plimpton 322 was first cataloged in 1943 for the Columbia University Library, where it is on permanent display, as just a "commercial account" such as an inventory or receipt, like hundreds of other recovered OB clay tablets. However, Otto Neugebauer uncannily saw that if he squared the value in Column III and subtracted from it the square of the value in Column II, he obtained a number

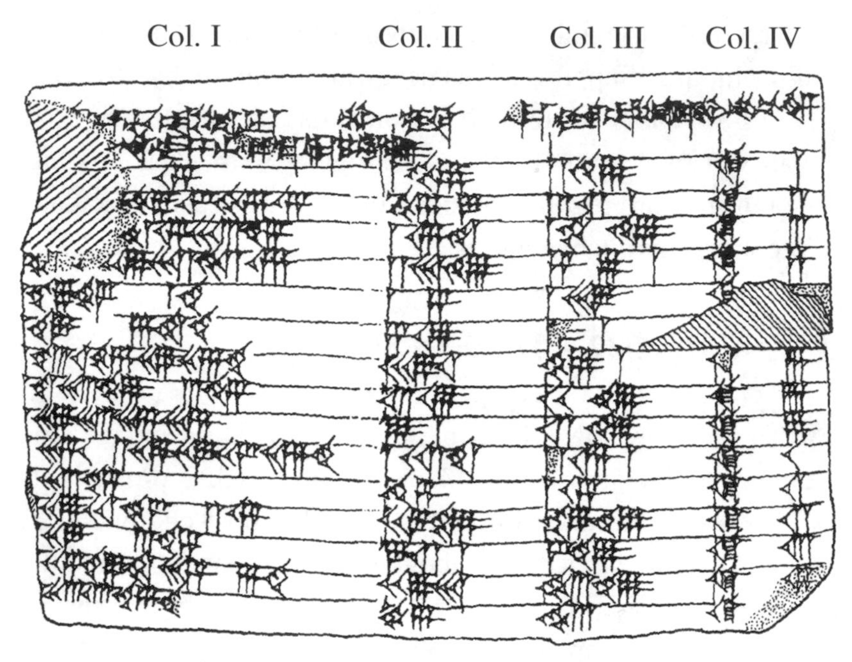

Figure 5.3.1 *Drawing of Plimpton 322*

that was itself a square of an integer. Neugebauer had discovered that Plimpton 322 was a table of *Pythagorean triples*, integer solutions to the *Pythagorean theorem*, $a^2 + b^2 = c^2$. At that moment, he converted a piece of clay, which had been purchased by George Plimpton in 1922 for $10, into a one-of-a-kind, OB mathematical table of incredible scholarly and monetary value. Without needing to understand a single cuneiform word, Neugebauer had conclusively proved that Plimpton 322 was about Pythagorean triples. But did it refer to right triangles?

Surprisingly perhaps, Pythagorean triples do not necessarily have anything to do with triangles. Figure 5.3.2 shows that $3^2 + 4^2 = 5^2$, the best-known Pythagorean triple, can be looked at as simply a relationship among three geometric squares, or in algebraic terms more abstractly as a relationship among three squared integers. In addition, as shown in the figure, triples are solutions to the Pythagorean theorem, $a^2 + b^2 = c^2$, which does relate to the sides of

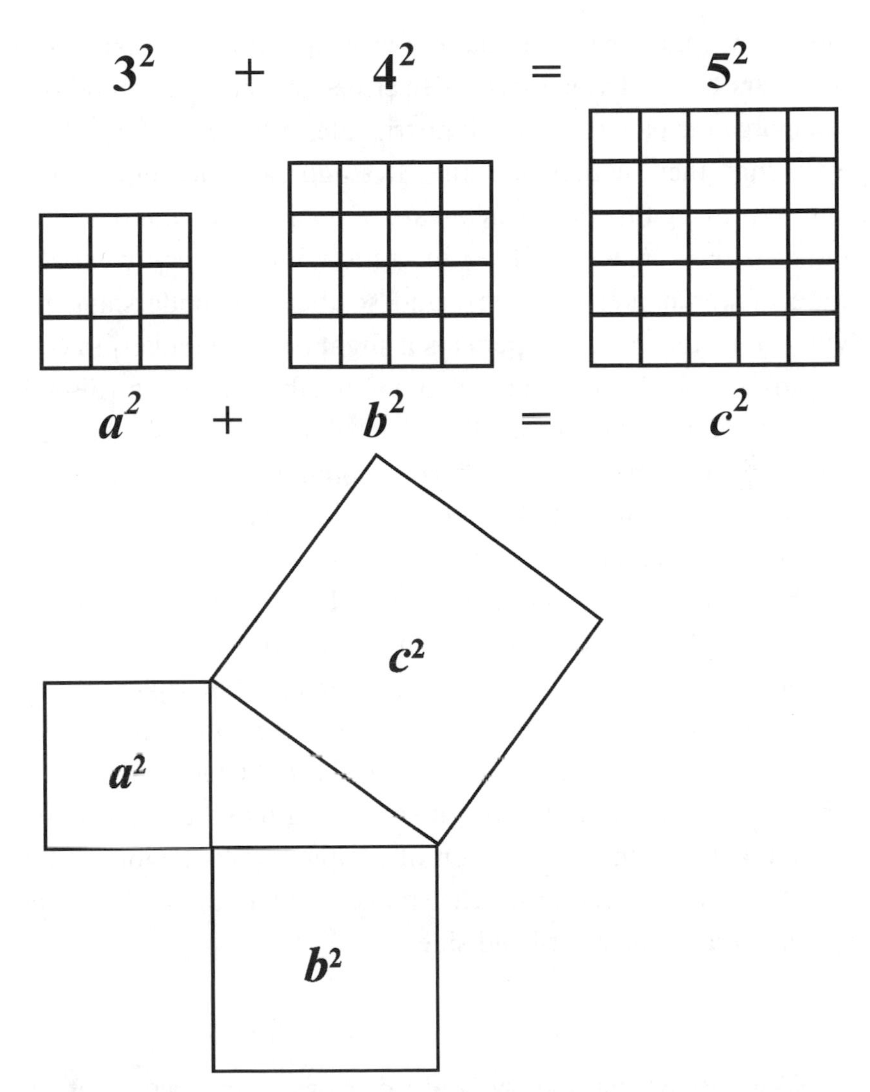

Figure 5.3.2 *Pythagorean triples and the Pythagorean theorem*

a right triangle. Every Pythagorean triple satisfies the Pythagorean theorem, but not every solution of the Pythagorean theorem is a triple. For example, $a = b = 1$, $c = \sqrt{2}$ satisfies the Pythagorean theorem, but $\sqrt{2}$ is not an integer and this (a, b, c) set is not a triple.

Neugebauer was also able to translate, in an incompletely translatable heading to Column III, the word "diagonal." With the translation of just that one word, he had proved that at least one thousand years before Pythagoras, OB scribes indeed understood that Pythagorean triples relate to right triangles. However, he did not prove that they understood the Pythagorean theorem, which requires proving that they were aware of noninteger solutions (a, b, c) to $a^2 + b^2 = c^2$. We shall see in section 5.5 that they were also aware of noninteger solutions, and so they did understand the Pythagorean theorem. It is perhaps a slight on Babylonians to continue to credit Pythagoras for a theorem that they and not he discovered, but that is still the generally accepted usage and I shall adhere to it. Probably some politically correct committee will soon recommend that we call it the Babylonian-Pythagorean theorem.

Why the Babylonians wanted *Pythagorean triples* is one mystery; *how* they derived these numbers is another. Let us begin with trying to find out how they arrived at the fifteen triples given in Plimpton 322.

The first general method for finding Pythagorean triples is credited to Diophantus of Alexandria (circa 250), more than fifteen hundred years after the writing of Plimpton 322. Let us look at how the Babylonians might have arrived at something like the Diophantine solution. I start with the definition of a triple, the Pythagorean theorem with only integer terms. Rearrange terms in the Pythagorean triple and *factor* the right-hand side:

$$b^2 = c^2 - a^2 = (c - a)(c + a)$$

Did the Babylonians know how to factor an algebraic equation? Remember, they did not have algebraic notation, just relationships between numbers and geometric visualizations. We now know from section 5.2 that they at least knew how to factor a regular number. In addition, they could have done so by the geometric algebra *of* figure 5.3.3, which is just figure 5.1.3 (with c replacing b), the visualization probably used to derive a practical multiplication algorithm. Now the visualization is not division of a square, but rather starting with a rec-

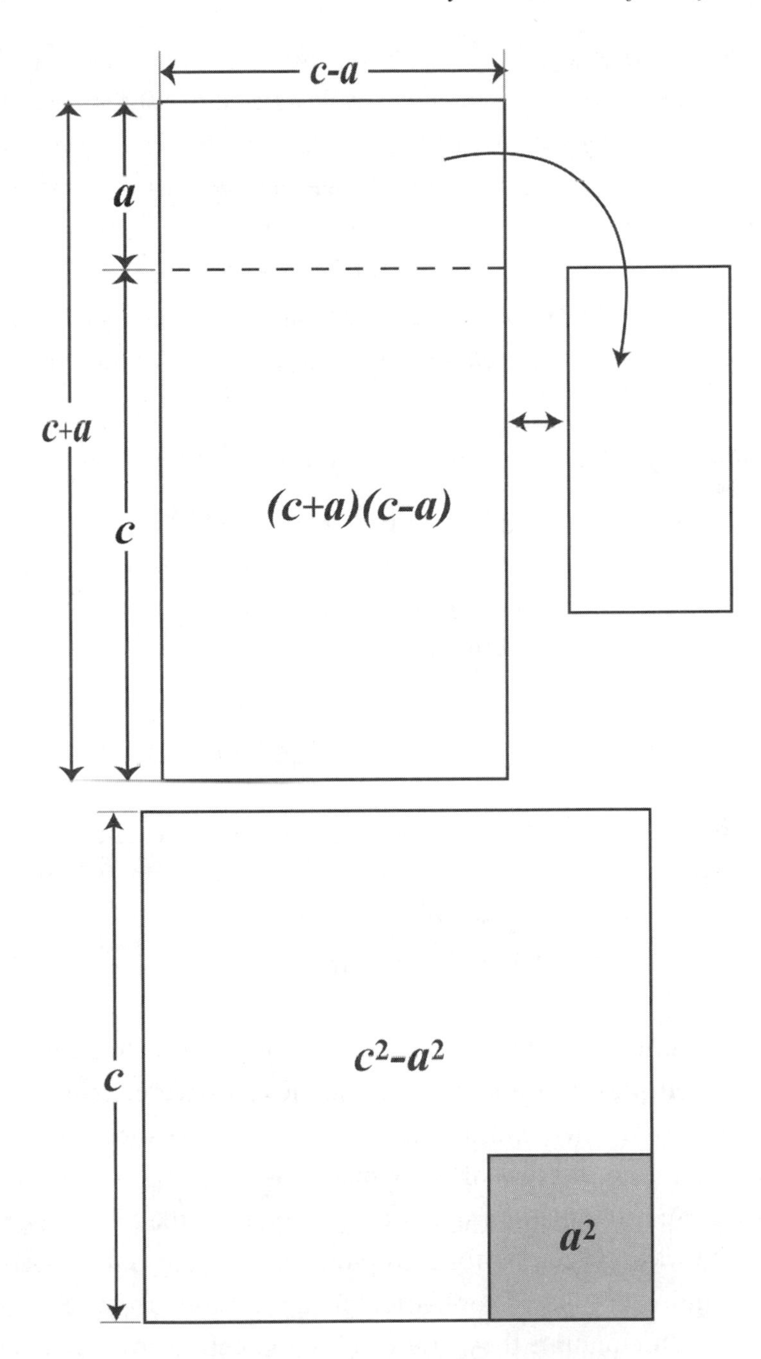

Figure 5.3.3 *Geometric algebra proof:* $c^2 - a^2 = (c + a)(c - a)$

tangle and cutting and pasting it to form a square. An area $a(c - a)$ is cut off from the $(c + a)(c - a)$ rectangle, right-rotated by 90 degrees, and pasted back, forming the inverted-L-shaped area that is now readily seen to be equal to $c^2 - a^2$. In section 5.4 we shall see that this cut-and-paste technique is apparently documented in surviving problem texts, so such OB use for such factoring is not too unreasonable.

Some insightful OB scribe might now have seen that by dividing the factored form of the Pythagorean theorem by b^2, he could obtain a reciprocal relation:

$$(c/b + a/b)(c/b - a/b) = 1, \text{ or equivalently, } (c/b + a/b) = 1/(c/b - a/b)$$

Since a, b, c must be integers, $(c/b + a/b)$ and $(c/b - a/b)$ must be common fractions and thus can be expressed as: $(c/b + a/b) = p/q$ and $(c/b - a/b) = q/p$, where p and q are also integers. Now by simple addition and subtraction:

$$[(c/b + a/b) + (c/b - a/b)]/2 = c/b = (p/q + q/p)/2$$
$$[(c/b + a/b) - (c/b - a/b)]/2 = a/b = (p/q - q/p)/2$$

Using these results in the Pythagorean theorem, we obtain it in *triples* form:

$$\begin{array}{ccccc} \mathbf{a^2} & \mathbf{+} & \mathbf{b^2} & \mathbf{=} & \mathbf{c^2} \\ \mathbf{(p^2 - q^2)^2} & \mathbf{+} & \mathbf{(2pq)^2} & \mathbf{=} & \mathbf{(p^2 + q^2)^2} \end{array} \qquad (5.3.1)$$

All that remains is to choose integers for p and q to generate Pythagorean triples. I doubt that the Babylonians were capable of such a derivation—my intuition is that it is just too sophisticated when compared to the rest of their mathematics. Yet, as we shall shortly see, this so-called pq method appears to be the best way, if not the only way, to account for Plimpton 322. But the Babylonians did not require any such sophisticated derivation because they already had an equation they could use to generate triples. They could have used the geometric algebra of figure 5.1.1, which was possibly their starting visualization for OB quadratic geometric algebra. In order to differentiate between *a, b* notation in equation (5.1.1) and *a, b* notation used in the Pythagorean theorem, since they do not refer to the same lengths, I rewrite equation (5.1.1) with a little rearrangement and change of symbols as:

$$(y - x)^2 + 4xy = (y + x)^2 \qquad (5.3.2)$$

If the term $4xy$ can be an integer squared, then this equation will be in Pythagorean triples form. We can convert the rectangle of area xy into a square of sides $\sqrt{xy}$. If we make $y = p^2$ and $x = q^2$, where p and q are integers, then $\sqrt{xy} = pq$, an integer, and $4xy = (2pq)^2$, an integer squared. Now equation (5.3.2) reproduces the Diophantine result, equation (5.3.1). It is also easy to see that there is also a more general solution: $y = mP^2$ and $x = mQ^2$, where m is any integer. This also reproduces the Diophantine result when used in equation (5.3.2). Thus, for every solution (P,Q) there is a more general solution (p,q) such that $p = P\sqrt{m}$ and $q = Q\sqrt{m}$.

FUN QUESTION 5.3.1: Two of the three numbers of a Pythagorean triple are known: 13 and 85. Find the third number.

Such a purely geometric-algebra derivation of the pq method for generating triples is more reasonable than essentially the Diophantine derivation. That OB mathematics had some similarities to the earliest Greek mathematics of Pythagoras is not too surprising, but Diophantus lived some 750 years later, after some 750 years of unprecedented mathematical progress. It is hard to believe that some OB scribe was capable of something akin to his sophisticated derivation. Now let us use the pq method to try and understand more about how the enigmatic Plimpton 322 was calculated.

Plimpton 322 is a trigonometric table, so we might expect it to be ordered, as trigonometric tables are intuitively ordered, according to angle. OB scribes used exactly the same concept of angle (slope) as the Egyptian *seked* (see section 4.4), thus *a/b* would define the angle and for *triples* the angle would be given as:

$$seked = a/b = (p^2 - q^2)/2pq = (p/q - q/p)/2 \qquad (5.3.3)$$

Calculation of the *seked* requires division by both *p* and *q*. To avoid nonterminating fractions, the nemesis of ancient and modern arithmetic, OB scribes limited choices for *p* and *q* to the sexagesimal regular numbers that appear in their tables of reciprocals, table 5.2.1.

p	q	p/q	Slope (degrees)	Col. I $(a/b)^2$ or $(c/b)^2$	$b=2pq$	Col.II $a=p^2-q^2$	Col.III $c=p^2+q^2$	Col. IV
12	5	2.4	44.8 or 45.2	(1).983	120	119	169	1
64	27	2.37	44.2 or 45.8	(1).949	3,456	3,367	4,825	2
75	32	2.34	43.8 or 46.2	(1).919	4,800	4,601	6,649	3
125	54	2.31	43.3 or 46.7	(1).886	13,500	12,709	18,514	4
9	4	2.25	42.1 or 47.9	(1).815	72	65	97	5
20	9	2.22	41.5 or 48.5	(1).785	360	319	481	6
54	25	2.16	40.3 or 49.7	(1).720	2,700	2,291	3,541	7
32	15	2.13	39.8 or 50.2	(1).692	960	799	1,249	8
25	12	2.08	38.7 or 51.3	(1).643	600	481	769	9
81	40	2.03	37.4 or 52.6	(1).586	6,480	4,961	8,161	10
2	1	2.00	36.9 or 53.1	(1).563	(60) 4	(45) 3	(75) 5	11
125	64	1.95	35.8 or 54.2	(1).519	16,000	11,529	19,721	
48	25	1.92	35.0 or 55.0	(1).489	2,400	1,679	2,929	12
15	8	1.88	33.9 or 56.1	(1).450	240	161	289	13
50	27	1.85	33.3 or 56.7	(1).430	2,700	1,771	3,229	14
9	5	1.8	31.9 or 58.1	(1).387	90	56	106	15

***Table 5.3.1** Plimpton 322:* pq *theory calculations— in base-10 numbers*

Table 5.3.1 exhibits the results of *pq* theory calculations, for all combinations of *p* and *q* in the OB table of reciprocals such that: $1.8 \le p/q \le 2.414$, and *p/q* is a reduced fraction (*p* and *q* have no common factors). The calculations almost perfectly account for

Plimpton 322. The grayed-out data are just to assist in understanding the calculations and do not appear on the tablet; nor do I imply that such data are in the broken-off section of the tablet. The row between tablet rows 11 and 12 is the only *pq* pair that was missed by whoever composed or copied the tablet. It is not an unreasonable error, considering that there are several other discrepancies between the *pq* calculation and the tablet data that have been explained as copying or calculation errors.

Only in tablet row 11 is there a major discrepancy. The Plimpton 322 data are in parentheses, and it can be seen that *a*, *b*, and *c* are all fifteen times the calculated values for the triple (3, 4, 5). This is consistent with the more general *pq* solution with $p = 4\sqrt{15}$ and $q = 3\sqrt{15}$, but it is the only case where *p* and *q* are not simply integers. I do not know why this discrepancy occurs and it is annoying, but rather than throw out the baby with the bathwater, I will stick with *pq* theory as presently the only reasonable explanation of how Plimpton 322 was calculated.

The *pq* calculations can be continued for all the remaining possible values, $0 < p/q < 1.8$ with *p* and $q \leq 125$, producing thirty-two more entries and thereby covering the angular range 0 to 90 degrees. It is possible that there were two more "Plimpton 322" tables, each also with sixteen rows, and with a little luck, they may yet be found, hopefully with each left side not broken off.

Plimpton 322 shows that OB scribes understood Pythagorean triples and perhaps the Pythagorean theorem. It also probably shows that an elegant solution, presumably the *pq* method, was derived. It also hints at some understanding of number concepts: *prime numbers, composite numbers, regular numbers, rational numbers,* and *reduced fractions*.

Now that we have some understanding about *how* Plimpton 322 was calculated, we can consider the even more speculative, but much more intriguing question of *why* it was calculated. While multiple copies of most OB tables have been found (multiplication tables, reciprocal tables, tables of squares, and so on), there is only one Plimpton 322. It is reasonable to guess that it was composed for some special purpose.

The generally accepted explanation is that Plimpton 322 was a "teacher's aid" for composing problems involving right triangles. If the tablet was to serve as a handy source for triples, then why is there no column containing values for b? The conventional answer is that there was, it just was in the broken-off part. However, the logical place for such a column is next to Columns II and III, as in table 5.3.1.

The most telling objection to the "teacher's aid" theory is that it does not explain the existence of Column I. In table 5.3.1 this column is labeled as either $(a/b)^2$ or $(c/b)^2$. Since the break in the tablet occurs just to the left of the column, there is a possibility that part of the column was cut off. If 1 was cut off, then the column values can be accounted for as $(c/b)^2$; if 1 was not cut off then the values can be accounted for as $(a/b)^2$. These alternatives are possible because of the Pythagorean theorem: $(c/b)^2 = 1 + (a/b)^2$. To me, the drawing of Plimpton 322 in figure 5.3.1 shows only a dubious hint that 1 has broken off. However, Neugebauer was able to translate one word in the Column I heading as "diagonal," the same word he found in the heading for Column III, and so the generally accepted interpretation is $(c/b)^2$. However, it does not make any practical difference whether $(a/b)^2$ or $(c/b)^2$ was on the undamaged tablet because whichever one was given, the other one could be readily obtained from it just by adding or subtracting 1.

If the "teacher's aid" explanation is rejected, what else could it be? What else is there that involves trigonometry? Construction. Construction of what? It probably would not have been just any building, but something special. The most sacred buildings in Babylon were the ziggurats, which were constructed from frustums

Figure 5.3.4 *Reconstruction of a Babylonian ziggurat*

of pyramids with square bases (see section 4.3). Figure 5.3.4 is a sketch of a model of a ziggurat based on archeological evidence.

The ziggurat in the city of Babylon was probably first built by Hammurabi and enlarged and reconstructed by Nebuchadnezzar into an awesome structure. During the Jewish exile in Babylon, following the conquest of Jerusalem by Nebuchadnezzar, there must have been a population that spoke a bewildering variety of languages reflecting the succession of different peoples who had occupied Babylon (Sumerians, Akkadians, Kassites, Assyrians, Persians, and so on). This combination of an awesome ziggurat in a city of many languages inspired some Jewish author to compose the biblical story of the Tower of Babel. In Hebrew, Babylon is *bbl* and the verb meaning to confuse is a similarly sounding *blbl*. This usage of the dual meaning of the word Babel is possibly the first recorded Jewish joke.

To find a historical basis for the speculation that Plimpton 322 relates to sacred mathematics, I return to ancient religious practice in India (see figure 3.2.2). A school of thought says that Indian mathematics originated in the service of religious ritual. My intuition is that Indian priests were simply awed by mathematical patterns, geometrical and numerical, and in turn used manipulation of these patterns to enhance their own stature and influence. (In section 6.1 we shall see that this also characterized the behavior of Pythagoras.) Surviving Hindu literature, known as the Sulbasutras (not nearly as well known as the Kamasutras), contains mathematical criteria for altar construction based on Pythagorean triples. Priest-scribes wrote down the Sulbasutras after 1000 BCE, but they presumably relate to orally transmitted traditions of religious practice that existed before 2000 BCE (see FUN QUESTION 4.1.2). The Babylonian culture is long deceased, and only an archeological record remains. Hindu culture survives, and its surviving literature provides a valuable clue to very ancient practice.

There is ample archeological evidence for contact and trade between Babylon and India, but I do not know whether the Indians learned the mathematics of the Sulbasutras from the Babylonians, or the Babylonians learned from the Indians. The Hindu texts are by no means proof that the construction of Babylonian ziggurats was according to the sacred mathematics of Pythagorean triples, but neither is this conjecture just rank speculation.

The Sulbasutras describe how to construct right triangles by stretching ropes. (See in section 3.1 the quotation from Ezekiel 40 for biblical reference to this technique. Interestingly, this quotation also relates to the sacred dimensions of a religious construction—the temple in Jerusalem.) The ropes were marked at intervals of a foot, and each side of the right triangle was an integral number of intervals. The Indians were clearly exploiting Pythagorean triples! What is intriguing about this rope-stretching use of Pythagorean triples is the implication that the discovery of Pythagorean triples probably preceded discovery of the Pythagorean theorem. Nowadays we tend to think that the Pythagorean theorem came first because we learn the Pythagorean theorem first and pay little attention to Pythagorean triples. However, four thousand years ago, there was not much practical use for the Pythagorean theorem because of the difficulty in calculating square roots, but certainly Pythagorean triples were used.

Ziggurats have sloping sides, so perhaps Plimpton 322 was used to define slopes. As illustrated in figure 5.3.5, every triangle defines two slopes, and my transcription of Plimpton 322 in table 5.3.1 lists both slopes. (For the benefit of those who understand some trigonometry, I calculated the angles as $\tan^{-1}[a/b]$ and $[90° - \tan^{-1}(a/b)]$.)

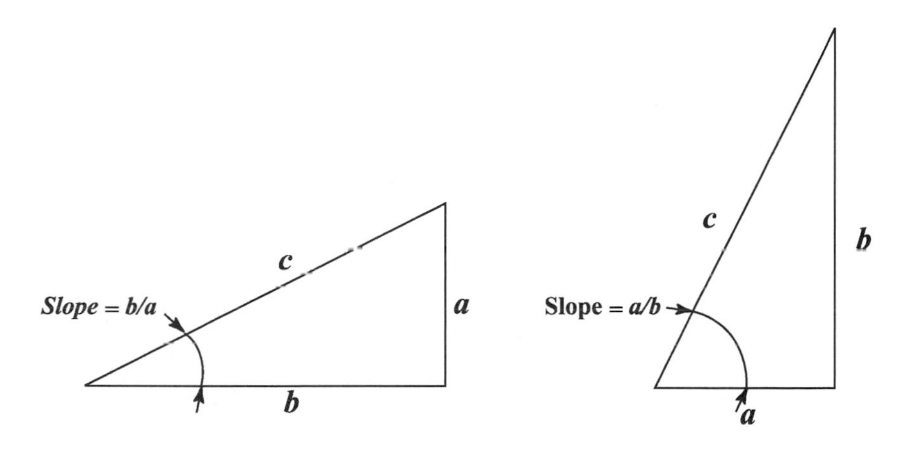

Figure 5.3.5 *Slopes defined by triangles*

Determining slopes may have been a use of Plimpton 322, but it doesn't account for Column I, which as just seen gives essentially both $(a/b)^2$ and $(c/b)^2$. To see why either of these values could have been important, consider the design of the Hindu sacrificial altar illustrated in figure 5.3.6. The triangles AXO and BYO, and their mirror images about the XY axis, DXO and CYO, determine the overall trapezoidal shape and dimensions of the altar. Triangles AXO and BYO are Pythagorean triples of dimensions (4×3, 4×4,

4×5) and (5×3, 5×4, 5×5), respectively. However, within the altar there are six more isosceles triangles that are constructed from Pythagorean triples, which are not based on the (3, 4, 5) triangle. There surely was ancient awareness outside of Babylon of Pythagorean triples other than just (3, 4, 5).

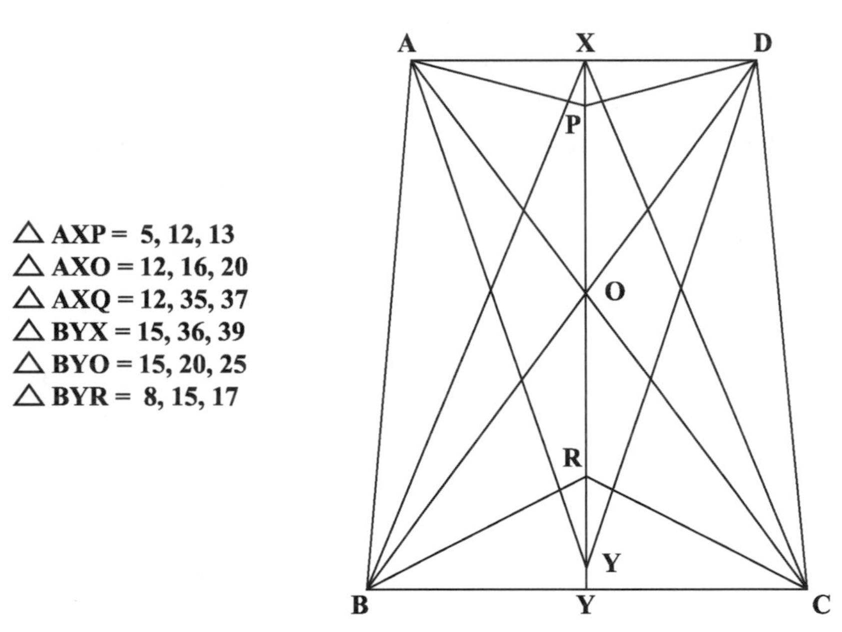

Figure 5.3.6 *Hindu sacrificial altar defined by Pythagorean triples*

I have no idea why it was so important to construct a sacrificial altar of exactly this shape and size. I can only note that the three inscribed isosceles triangles pointing up, and the three pointing down, probably have the same symbolic meaning as previously noted with respect to figure 3.2.2.

Not only can rope stretching be used to construct a right triangle, it can also be used as a device for finding Pythagorean triples: for different lengths of rope, see if a right triangle can be formed with integral sides. For the triples of figure 5.3.6, this could have been the method. In principle, it could even have been used to find the triples of Plimpton 322. However, from table 5.3.1, when we note the long lengths required to define most of the triples, it is an unlikely method.

Observing that the sides of a frustum and hence of a ziggurat are trapezoids, we have another way in which Pythagorean triples could have played a role in ziggurat design in a manner analogous to Hindu altar design. For example, frustums with easily calculable ratios of top areas to bottom areas can be constructed if trapezoids are constructed from triples as illustrated in figure 5.3.7 for $(a, b, c) = (3, 4, 5)$. The area ratios can be either $(a/b)^2$ or $(c/b)^2$, both of which are essentially given in Column I of Plimpton 322 and have the convenience of being terminating fractions, or $(a/c)^2$. A designer of a ziggurat would have wanted to know such a ratio. By limiting use to the Plimpton 322 tabulation of Pythagorean triples, the designer largely obviated any need to calculate square roots, which he probably did not know how to do or at best only by a very tedious calculation (see section 5.5).

FUN QUESTION 5.3.2: If the Pythagorean triple is (3, 4, 5), what is the vertical distance, h, between the base and the top of the frustum with a top-to-bottom ratio of $(a/b)^2$ of figure 5.3.7? What is the *seked* and the angle of the slope for this frustum? Some trigonometry is required.

I have not proved *why* Plimpton 322 was composed. However, I have shown that as calculations employed in ziggurat construction, its purpose is perhaps better explained than by the generally accepted "teacher's aid" hypothesis. The *why* of Plimpton 322 has not been solved, and unless other relevant clay tablets are unearthed, it possibly never will be.

The mystery of Pythagorean triples does not end with the rigorous treatment given by Diophantus of Alexandria around the year

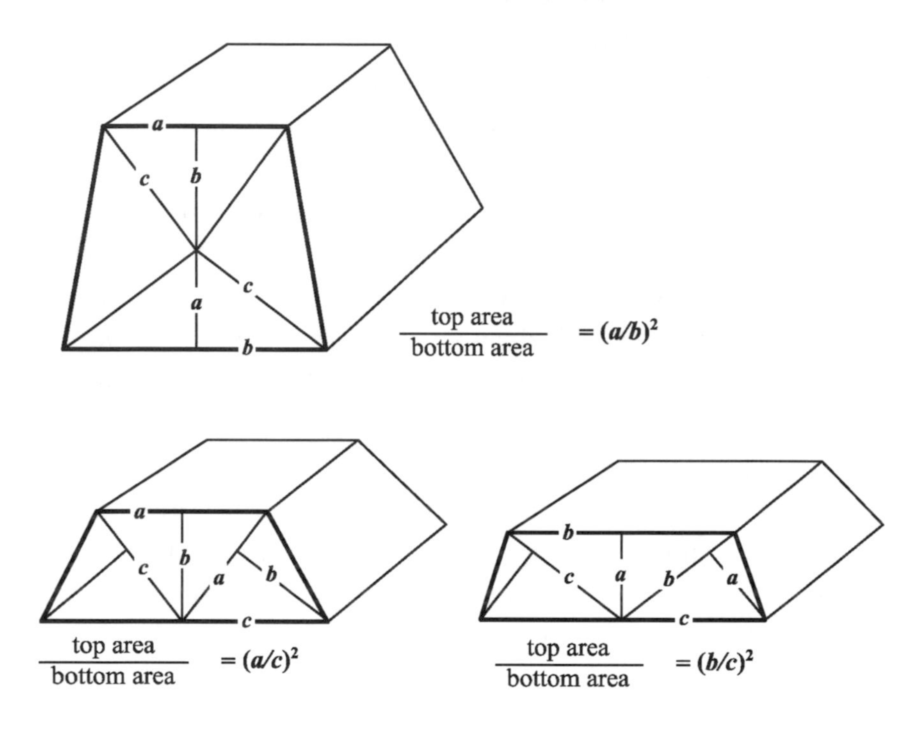

Figure 5.3.7 *Frustum defined by Pythagorean triples (*a,b,c*)*

250 in his book *Arithmetic*. A copy of this book miraculously survived the destruction of the Library of Alexandria in 389, when Christians trashed pagan monuments, and the destruction in 642, when Muslims trashed any book that contradicted the Koran as well as any that agreed because that meant they were superfluous. In 1638 a copy of *Arithmetic* inspired the mathematician Pierre de Fermat to state that he had proved, in a short proof that wouldn't quite fit in the margins of the book, that there was no solution to a *generalized Pythagorean triples* problem: $a^n + b^n = c^n$ (a, b, c, and

n are integers and $n > 2$). This claim, now known as *Fermat's last theorem*, has confounded the greatest of mathematicians.

It was only in 1995 that Andrew Wiles, of the Department of Mathematics, Princeton University, using the most sophisticated modern techniques, in an article of more than one hundred pages, succeeded in proving that Fermat's last theorem is indeed correct. It is now clear that Fermat's proof was either wrong or that he knew that there was no simple proof and he was just tormenting his successor mathematicians. At least this problem with four thousand-year-old roots has finally been solved, but the enigma of Plimpton 322 remains to keep us entertained.

5.4 Babylonian Algebra

Clay tablets known as *problem texts* provide evidence of OB geometric algebra. A typical OB problem text is YBC 6967 (YBC = Yale Babylonian Collection): The length of a rectangle exceeds its width by 7_{60}. Its area is $1{:}0_{60}$. Find its length and width. Expressed in modern algebraic notation: $b - a = 7_{60}$, $ab = 1{:}0_{60} = 60_{10}$.

	Babylonian algorithm	**Algebraic generalization**
1.	$7/2 = 3.30$	$(b - a)/2$
2.	$3.30 \times 3.30 = 12.15$	$[(b - a)/2]^2$
3.	$12.15 + 1{:}0. = 1{:}12.15$	$[(b + a)/2]^2 = [(b - \mathrm{a})/2]^2 + ab$
4.	$\sqrt{1{:}12.15} = 8.30$	$(b = a)/2 = \sqrt{[(b-a)/2]^2 + ab}$
5.	$8.30 - 3.30 = 5 =$ width	$a = (b + a)/2 - (b - a)/2$
6.	$8.30 + 3.30 = 12 =$ length	$b = (b + \mathrm{a})/2 + (b - a)/2$

This algorithm can be derived by *geometric algebra* from figure 5.4.1, which is simply figure 5.1.1 with all dimensions halved. This

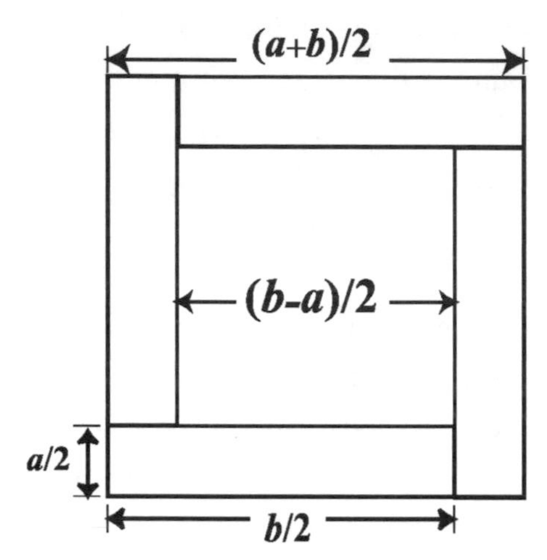

Figure 5.4.1 *Geometric-algebra visualization of problem text YBC 6967*

diagram relates to equation (5.1.1a), which is an equation that probably was used for actual multiplication using tables of squares.

Figure 5.4.1 defines a relationship between three areas: ab, $[(b - a)/2]^2$, and $[(b + a)/2]^2$. The only difference between the algorithm for solution of this problem text and the multiplication algorithm using tables of squares is that in YBC 6967 ab and $(b - a)$ are the givens and $(b + a)$ is the unknown, while in the multiplication algorithm $(b - a)$ and $(b + a)$ are the givens and ab is the *unknown*.

Step 1 calculates $(b - a)/2$; **Step 2** calculates the second area, $[(b - a)/2]^2$; **Step 3** calculates the third and unknown area $[(b + a)/2]^2$. **Step 4** takes the square root of this area to obtain $(b + a)/2$. Now both $(b - a)/2$ and $(b + a)/2$ are known and the unknown sides (a and b) are easily calculated as sums and differences of these quantities in **Steps 5** and **6**.

Step 4 requires taking a square root of a number that is not simply an integer squared; that is not a problem that had been generally solved during the childhood of mathematics. In fact, pencil/paper calculation of such a square root would be a problem for most people even today. Only a couple of generations ago, calculation of square roots was part of elementary arithmetic instruction, but the ubiquitous electronic calculator has now rendered such calculation obsolete (see section 7.1 for more on pencil/paper calculation of square roots). The teacher-scribe who composed the problem of YBC 6967 did not have to know how to calculate the square root because he composed the problem by setting $a = 5$ and $b = 12$, and hence he could have simply back-calculated the square root as $(a + b)/2$. Assuming that one of the purposes of contriving the problem YBC 6967 was to teach at least something about how to cope with taking square roots, let us consider how a Babylonian student-scribe might have calculated $\sqrt{72.25}$ ($1{:}12.15_{60} = 72.25_{10}$).

One possible method could have been simply by successive guessing: since $9^2 > 72.25$ and $8^2 < 72.25$, try 8.5^2, and that works (here and in the remainder of this discussion about calculation of square roots I shall use decimal numbers rather than decimally transcribed sexagesimal numbers to make the arithmetic easier).

It is also possible that the square root calculation was another exercise in geometric-algebra visualization. Let us consider the calculation first in modern algebraic notation. A possible solution is $72.25 = (j + 1/2)^2$, where j is an integer, because it is obvious that $(1/2)^2$ is required to account for 0.25. Algebraically expanding the squared term we obtain $72 = j^2 + j$. Now from a table of squares of integers, it is easy to see that the solution is $j = 8$. This solution can be visualized by the construction of figure 5.4.2 that is very similar to the visualization used to derive the YBC 6967 algorithm. I present this visualization because, as we shall see later in section 5.5, OB scribes did eventually invent a general method for calculating square roots that is just a variation on the construction of figure 5.4.2, and so this visualization is a logical predecessor invention. Although OB scribes had a completely general square root solution, they did not completely understand their solution and apparently only correctly applied it to calculate $\sqrt{2}$.

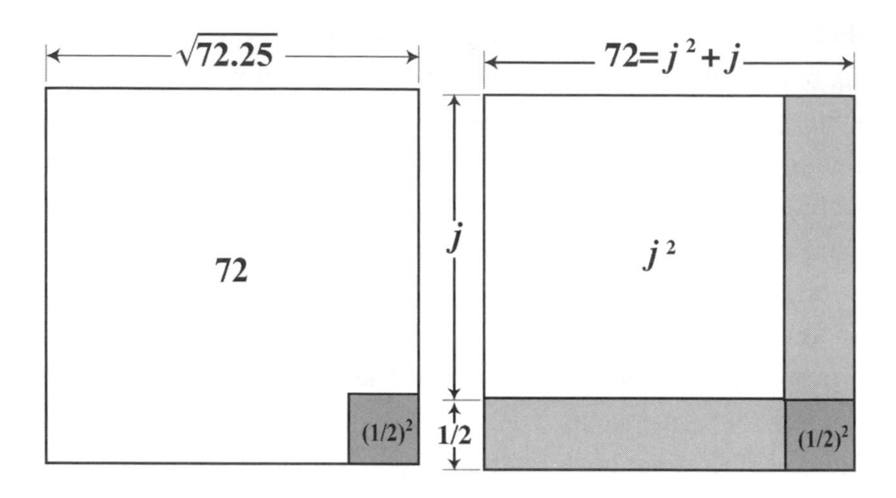

***Figure 5.4.2** Geometric-algebra visualization for calculating a square root*

Another way to calculate this square root is to convert 72.25 into a common fraction: 72.25 = 289/4, and from a *table text* table of squares to have seen that $289/4 = (17/2)^2$. In every case where a problem text requires taking a square root, the use of a table of squares of integers can solve the problem.

YBC 6967 is a contrived problem to teach geometric-algebra visualization to student-scribes. It is unlikely that any practical problem inspired the composition of this problem. The solutions, $a = 5$, $b = 12$, are the sides of a rectangle whose diagonal is 13, and a (5, 12, 13) triangle is a Pythagorean triple. In this particular problem the diagonal plays no role, but this same rectangle was also used in other problems where the diagonal was a variable. The problems were contrived never to encounter nonterminating fractions.

This OB *problem text* is very similar to the type of contrived problems that appear in present-day algebra texts: "The product of the ages of Pat and Mike is 60. Pat is seven years older than Mike. How old are Pat and Mike?" The present-day role of such contrived problems is to teach the distilling of a rhetorical expression into an algebraic equation for which there is a general solution. The Babylonian role for their contrived problems was similar, to teach distilling of a rhetorical expression into a geometric construction for which there was a general solution. Present-day high school students generally hate such word problems because they think they are just contrived to torture. Babylonian student-scribes probably felt the same way. Teachers seldom take the trouble to explain the purpose of such word problems, although I am not sure it would make much difference if they did.

In writing the YBC 6967 algorithm, as previously with the Egyptian frustum algorithm (see section 4.3), I have placed a minimal burden on the translation of the text; the numbers speak for themselves. However, recent progress in translating OB cuneiform algorithms leads to an interpretation that not only are they derived from geometric diagrams, but also that the algorithms can sometimes be read as a prescription for cut-and-paste construction of a diagram. Figure 5.4.3 is the construction proposed for YBC 6967.

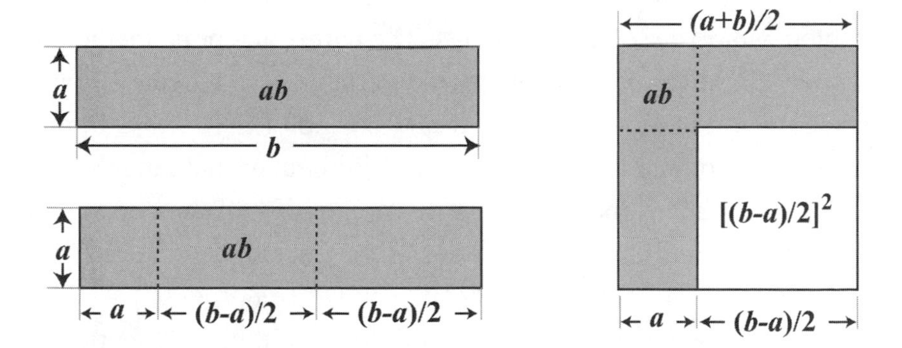

Figure 5.4.3 *Cut-and-paste interpretation of YBC 6967 algorithm*

The rectangle of area *ab* is cut and pasted to form a *gnomon* of the same area. (The word *gnomon* is the Greek word for the pointer that casts a shadow on a sundial. Ancient sundials had an "inverted L-shaped" pointer. In current geometry jargon the word *gnomon* defines an L-shaped area, and it is still used as the name of the pointer on a sundial, whatever its shape.) We have previously used this geometric construction in figure 5.3.3 to factor the Pythagorean theorem, and figure 5.4.3 is but the same figure redimensioned to relate to YBC 6967. Figures 5.4.1 and 5.4.3 both have exactly the same dimensions, so both account equally well for the algorithm as I have transcribed it.

Adding to the *gnomon* a square of area $[(b - a)/2]^2$ creates a square of area $[(a + b)/2]^2$. The *"cut-and-paste"* solution is literally by *completing the square*. Whether the OB algorithm relates to figure 5.4.1 or 5.4.3, I will leave for translators of cuneiform to thrash out. It is nice to have this new translation confirm that OB scribes were doing geometric algebra, but it is not essential. With or without cut-and-paste solutions, it is now obvious that Otto Neugebauer misinterpreted problem-text algorithms as purely algebraic derivations (see section 5.1).

It is interesting to compare the OB solution with how this problem would be solved nowadays. The purely algebraic method of solving YBC 6967 is to eliminate the variable b and create a quadratic equation in the variable a: $a(a + 7) = 60$ or $a^2 + 7a = 60$, and solve it by completing the square, but without visualizing a geometric square. This means adding the square of half the coefficient of a, $(7/2)^2$ in this case, to both sides of the quadratic equation: $a^2 + 7a + (7/2)^2 = [a + 7/2]^2 = 60 + (7/2)^2$. Thus $a = \sqrt{60 + (7/2)^2} - 7/2 = 5$, and $b = a + 7 = 12$. Step-for-step, the algorithm for this purely algebraic solution is not identical to the the geometric-algebra algorithm, but the concept of completing the square is the same.

Another OB problem text that illustrates more complex geometric algebra is a more recent find, in 1962 at Tel-Dhibayi, the ruins of ancient Eshnunna, and is designated as Db_2 146: The area of a rectangle is 0.45_{60} and its diagonal is 1.15_{60}. Find its length and width. Expressed in modern algebraic notation: $ab = 0.45_{60}$, $c = 1.15_{60}$, find a and b.

Babylonian algorithm	**Algebraic generalization**
1. $2 \times 0.45 = 1.30$	$2ab$
2. $(1.15)^2 = 1.33{:}45$	c^2
3. $1.33{:}45 - 1.30 = 0.3{:}45$	$(b - a)^2 = c^2 - 2ab$
4. $\sqrt{0.3:45} = 0.15$	$(b - a) = \sqrt{c^2 - 2ab}$
5. $0.5/2 = 0.7{:}30.$	$(b - a)/2 = \sqrt{c^2 - 2ab}/2$
6. $0.3{:}45/4 = 0.0{:}56{:}15$	$[(b - a)/2]^2 = (c^2 - 2ab)/4$
7. $0.45 + 0.0{:}56{:}15 =$ $0.45{:}56{:}15$	$[(a + b)/2]^2 = (c^2 - 2ab)/4 + ab = (c^2 + 2ab)/4$
8. $\sqrt{0.45:56:15} = 0.52{:}30$	$(a + b)/2 = \sqrt{c^2 + 2ab}/2$
9. $0.52{:}30 + 0.7{:}30 = 1.0$	$b = (a + b)/2 + (b - a)/2$
10. $0.52{:}30 - 0.7{:}30 = 0.45$	$a = (a + b)/2 - (b - a)/2$

For an algorithm of so many steps, it is difficult to conceive that the solution could have been mastered without reference to a geometric diagram as a mnemonic. A geometric-algebra visualization for this algorithm is the diagram of figure 5.4.4. It is exactly figure 5.1.1, but with a diagonal drawn through each rectangle. The square of area c^2 is composed of four triangles (shaded) of area $2ab$ and a square (shaded) of area $(b - a)^2$. **Steps 1–5** use this geometry to calculate $(b - a)/2$ from the givens, c and ab. Additionally, the square of area $(a + b)^2$ is composed of four triangles of area $2ab$ (not shaded) and a square of area c^2. **Steps 6–8** use this geometry to calculate $(a + b)/2$. **Steps 9** and **10** calculate a and b just as in YBC 6967.

Steps 4 and **8** may appear to have been formidable square root calculations, but in fact they could have been simply obtained by converting to common fractions and using a table of squares of integers: $\sqrt{0.3:45} = \sqrt{1/16} = 1/4$ and $\sqrt{0.45:56:15} = \sqrt{49/64} =$

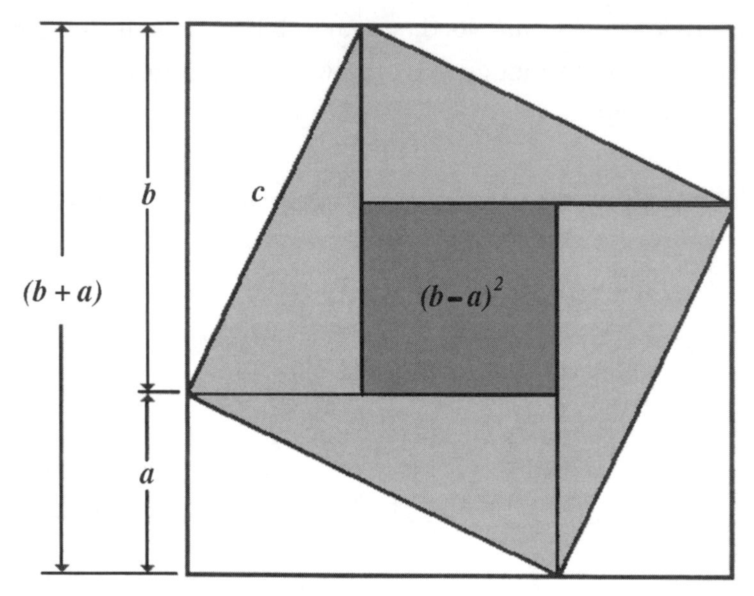

Figure 5.4.4. *Geometric-algebra visualization of problem text* Db_2 *146*

7/8. As noted previously, there is little documentary evidence for intermediate arithmetic steps, and so it is not known how OB scribes did these square roots.

Although Db_2 146 involves the diagonal of a rectangle, the Pythagorean theorem was not invoked in the solution. What is particularly intriguing about this geometric-algebra solution is that it provides the diagrams for two known proofs of the Pythagorean theorem. From **Step 7** of the algorithm we have $(a + b)^2 = c^2 + 2ab$. From figure 5.1.2 we previously obtained $(a + b)^2 = a^2 + b^2 + 2ab$; equating these two expressions for $(a + b)^2$, we obtain $a^2 + b^2 = c^2$, the Pythagorean theorem. This is a proof of the Pythagorean theorem and not just of Pythagorean triples because nowhere is it required that a, b, and c be integers, even though the actual numbers in the problem do describe Pythagorean triples. Legend credits this proof to Pythagoras himself.

From **Step 3** we have $(b - a)^2 = c^2 - 2ab$. Figure 5.1.3 previously gave $(b - a)^2 = a^2 + b^2 - 2ab$; equating these two expressions for $(b - a)^2$, we again obtain $a^2 + b^2 = c^2$, the Pythagorean theorem. This proof is attributed to the Indian mathematician Bhaskara (circa 1150), although it could quite possibly have *diffused* from Babylon many centuries earlier. Did the Babylonians recognize the proofs inherent in this diagram and were they not only aware of the Pythagorean theorem, but possibly had also proved it?

Derivation of the Pythagorean theorem has been a continuing mathematical recreation for hundreds of generations, and there are hundreds of proofs. Even James Garfield, the twentieth president of the United States, published a proof. Prior to becoming president, Garfield had been a college professor of Latin and Greek, so his academic credentials were somewhat more than most politicians can claim. Only one hundred days after his inauguration, Garfield was assassinated, and so he never really had a chance to show his presidential talents, mathematical or otherwise.

FUN QUESTION 5.4.1: Prove that for a right triangle when $a/b = 1$, $p/q = 1+\sqrt{2}$. Prove this by making the equation (5.3.3) for the *seked* of a triangle (a, a, c) into a quadratic equation in a variable $r = p/q$ and solve the equation by cut-and-paste geometric algebra.

FUN QUESTION 5.4.2: The Greek mathematician Eudoxos (408–355 BCE) was abstractly interested in proportions and considered a rectangle ab proportioned as $a/b = b/(a + b)$. Somehow, Eudoxos was able to give his result the "spin" that it epitomized aesthetic symmetry. For almost twenty-five hundred years buildings, doors, picture frames, and various other constructions have been proportioned according to this *golden section,* although in recent centuries it has been recognized for the nonsense that it is. Convert Eudoxos's proportion into a quadratic equation in a variable $r = a/b$. Solve the equation by cut-and-paste geometric algebra and show that the golden section is given as: $a/b = (\sqrt{5} - 1)/2$.

FUN QUESTION 5.4.3: Show that YBC 6967 defines the triple (5, 12, 13). Why does this triple not appear on Plimpton 322? Show that Db_2 146 defines the triple (3, 4, 5).

5.5 BABYLONIAN CALCULATION OF SQUARE ROOT OF 2

The case for OB use of geometric algebra is strong, but the evidence presented so far for an OB proof of the Pythagorean theorem, or even an awareness of anything except Pythagorean triples, is weak. The dearth of archeological artifacts with diagrams that unambiguously relate to the Pythagorean theorem casts some doubt on Babylonian appreciation of the theorem. But neither is their such evidence for the diagrams attributed to Pythagoras, because he apparently did his diagrams by arranging pebbles or by drawing traces in sand, as probably did OB scribes. This scarcity of more archeological evidence is not fatal, just a little uncomfortable.

However, a surviving OB tablet, designated YBC 7289, exhibits a drawing that lays to rest most doubts concerning Babylonian understanding of the Pythagorean theorem. Figure 5.5.1 is a drawing of the tablet and a transcription of the cuneiform into base-10 symbols. This palm-sized tablet, which has become so critical to appreciating OB mathematics, is apparently just the notes of a student-scribe.

Clearly inscribed on the clay tablet, after making a reasonable assumption about where to put the *sexagesimal point* and decimally transcribing it, is the number 1.24:51:10, which when translated into a base-10 number is 1.41421297. The correct base-10 value of the square root of 2, to eight-digit precision, is 1.4142136. The Baby-

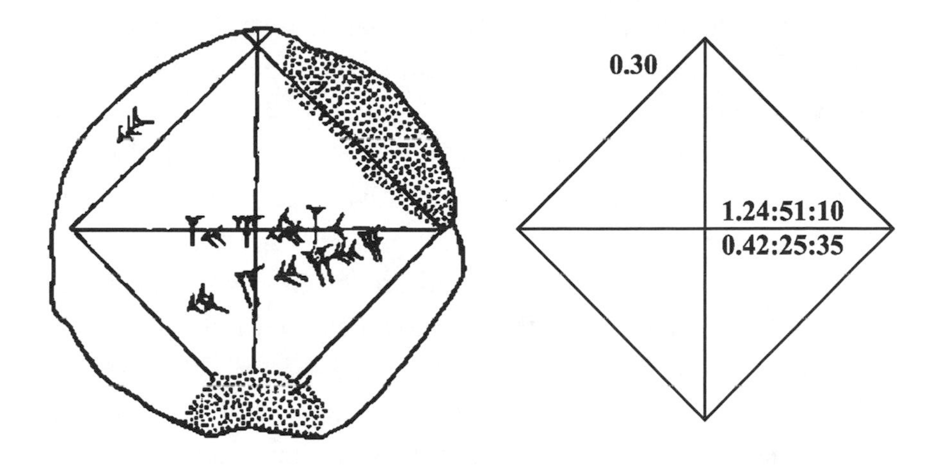

Figure 5.5.1 *Babylonian clay tablet YBC 7289: square root of 2*

lonian calculation is presumably of $\sqrt{2}$, correct to a remarkable seven-digit precision. This calculation relates to the right triangle formed by the drawn square and its diagonal, $(a, a, a\sqrt{2})$; with $a = 1/2$, the diagonal is $\sqrt{2}/2 = 1/\sqrt{2}$. The decimally transcribed number written just below the diagonal is 0.42:25:35, which transcribes into the base-10 number 0.7069162; the correct base-10 value for $1/\sqrt{2}$, to seven-digit precision, is 0.7071067. The internal consistency of all the numbers on YBC 7289 gives confidence that the square root of 2 has really been calculated and that it is not just a coincidence—as concluded previously about $\sqrt{3}$ appearing in calculations about pyramid dimensions (see section 4.4).

Not only does YBC 7289 show that the Babylonians had an algorithm to calculate $\sqrt{2}$ with surprising precision, but it is also a proof of the Pythagorean theorem for the special case of $a = b$. The diagram on YBC 7289 is exactly the proof credited to Bhaskara in Figure 5.4.3, but with $a = b$. The area, c^2, of the square in YBC 7289 is now equal to the area of four triangles, each of area $a^2/2$. This proves that for these right triangles $2a^2 = c^2$, which is the

Pythagorean theorem. This is a Babylonian artifact that unambiguously exhibits a proof of the Pythagorean theorem, albeit only for the specific case of $a = b$.

Scholars of OB mathematics now tend to believe that OB scribes not only used, but also had proved the Pythagorean theorem. In section 4.3 (see figure 4.3.4), it was speculated that the Egyptians also could have used and proved the Pythagorean theorem, but the evidence for this to have occurred is much weaker than for Babylon.

How did OB scribes calculate a square root (at least of 2) to such remarkable precision? A generally accepted answer is that they used a method credited to Heron of Alexandria (circa 150), some two thousand years after the presumed Babylonian usage. Heron's algorithm is one of successively improving approximations. The precision of the result only depends on the number of iterations. Let us consider Heron's algorithm in some detail in order to judge whether its prior invention by the Babylonians is a reasonable conjecture. We consider a right triangle $(a, a, c) = (1, 1, \sqrt{2})$.

Let c_0 be the first guess for the value of c. Because $\sqrt{1} = 1$, $\sqrt{2} =$ c, and $\sqrt{4} = 2$, clearly $1 < c < 2$, so let's take $c_0 = 1.5$. Since $c_0^2 = (1.5)^2 = 2.25 > 2$, we know that $c_0 > \sqrt{2}$, and therefore $c^2/c_0 = 2/c_0 < \sqrt{2}$. The average, $c_1 = (c_0 + c^2/c_0)/2 = (c_0 + 2\ /c_0)/2$, is a better approximation. The beauty of this algorithm is that the error in c_0 in one direction closely compensates the error in c^2/c_0 in the opposite direction, so c_1 is a much better approximation than c_0. To see why this error compensation is so good, consider table 5.5.1:

ε	$1 + \varepsilon$	$1/(1 + \varepsilon)$	$1 - \varepsilon$
0.2	1.2	0.83	0.80
0.1	1.1	0.91	0.90
0.01	1.01	0.9901	0.99

Table 5.5.1 *Small number approximation*

(ε = Greek letter epsilon, frequently used in mathematics to represent a small number.)

We see that as $\varepsilon \to 0$ (read: as ε approaches zero) $1/(1 + \varepsilon) \to 1 - \varepsilon$, and we would similarly find that $1/(1 - \varepsilon) \to 1 + \varepsilon$. If we write $c_0 = c(1 + \varepsilon)$, then $c_1 = c_0 + c^2/c_0 \cong c[(1 + \varepsilon) + (1 - \varepsilon)]/2 = c$. Thus, the major part of the error cancels out, which produces a much better approximation for c.

This is a very subtle step. I doubt that it was within OB mathematical competence. In addition, we have seen that OB algebra relied on geometric constructions, which are missing here. Even though Heron's derivation was probably not used, let us see how Heron's algorithm works:

Using our first guess of $c_0 = 1.5$, we obtain a better approximation:

$$c_1 = (c_0 + 2/c_0)/2 = (1.5 + 2/1.5)/2 = 1.4166667$$

In a similar way, we can obtain a more precise approximation:

$$c_2 = (c_1 + 2/c_1)/2 = (1.4166667 + 2/1.4166667)/2 = 1.4142157$$

A third iteration yields a result that is precise to eight digits:

$$c_3 = (c_2 + 2/_2)/2 = (1.4142157 + 2/1.4142157)/2 = 1.4142136$$

The method actually used by the OB scribes to calculate square roots was another iterative algorithm sometimes credited to Sir Isaac Newton (1642–1727), more than three thousand years after presumed Babylonian usage. Again, c_0 is the first guess, but now an insightful new concept is introduced: e_0 is the error, the amount that c_0 differs from $\sqrt{2}$, so $c_0 + e_0 = c$ and $(c_0 + e_0)^2 = 2$. It is also more reasonable that OB scribes used this method, because it can be derived from a geometric-algebra diagram with which they were familiar. We first encountered this diagram in figure 5.1.2 to derive a multiplication algorithm, practical arithmetic that was probably a starting point of OB awareness of geometric algebra. We then

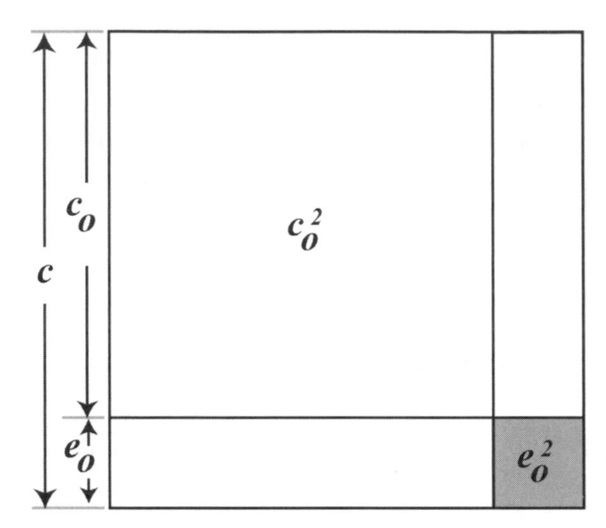

***Figure 5.5.2** Geometric-algebra derivation of OB, square root calculation*

encountered it again in figure 5.4.1 to solve a problem text used to teach geometric-algebra visualization. We have also seen how it could have been applied to calculate a square root in figure 5.4.2. Now we shall see how the geometric-algebra visualization learned in the problem texts was used to produce new practical arithmetic, a general solution to the square root problem.

Referring to figure 5.5.2: $(c_0 + e_0)^2 = c_0^2 + 2c_0e_0 + e_0^2 = 2$. If we have made a good choice for c_0, then $e_0 \ll c_0$ and the term e_0^2 is much smaller than the other terms, and a reasonable approximation is then that $c_0^2 + 2c_0e_0 \cong 2$, so that $e_0 \cong (2 - c_0^2)/ 2c_0$. Thus $c_1 = c_0 + e_0 \cong (c_0 + 2/c_0)/2$ is a better approximation and it is exactly the same as the result obtained by Heron's method. Thus, it really does not make much difference whether it is called Heron's or Newton's method because the end result is the same.

FUN QUESTION 5.5.1: Calculate $\sqrt{5}$ using three iterations of Heron's algorithm. This problem illustrates that, when properly applied, Heron's algorithm can calculate the square root of any number. However, the Babylonians apparently did not fully understand how to use their solution and only correctly applied it to calculate $\sqrt{2}$.

---

*   *   *   *   *

The story of the deciphering of Babylonian cuneiform parallels the "Rosetta Stone story" to a remarkable degree. In the 1830s, Henry Rawlinson, then an officer in the British army, learned of a large cuneiform inscription high up on a cliff of a mountain called Behistun in present-day Iran, some 250 km east of Baghdad. With great risk to his life, Rawlinson scaled the cliff, some 100 meters above ground level, and suspended from ropes succeeded in copying most of this *Behistun Inscription*. Figure 5.5.3 is a photograph of the Behistun Inscription.

The inscription had been commissioned by Darius I of Persia in 519 BCE and gave his boastful account of how he had come to the throne of the Persian empire. On completion of the inscription, a worker-ledge was removed so that nobody could tamper with it, and thanks to this it has remained readable; but neither could it be read by anybody, so for whose eyes had Darius written it? Obviously, for Henry Rawlinson's.

The same story had been inscribed in three different types of cuneiform writing: Old Persian, Babylonian, and Elamite. The Old Persian symbols were the simplest and, with its similarities to contemporary Persian dialects, Rawlinson succeeded in deciphering this part of the inscription by 1846. Then with the aid of his deciphered Persian text, Rawlinson completed deciphering the more difficult Babylonian cuneiform by 1857.

However, it was not until the publication in 1945 of Otto Neugebauer's virtuoso *Mathematical Cuneiform Texts* that a translation and analysis of Babylonian mathematics tablets became

Figure 5.5.3 *The Behistun Inscription*

widely available. I can only marvel at his having been able to decipher algebraic content from a jumble of barely identifiable indentations in chunks of clay. But he did make mistakes. He misinterpreted *generation* of algebra as *derivation* of algebra. He was also content to have algebra without antecedents. We have now seen how geometric algebra explains the derivation of Babylonian algorithms and how the quadratic-algebra problem texts probably evolved as generalizations of antecedent practical solutions to their unique problem of multiplication with a large base. We have also seen how mastering quadratic geometric algebra with the help of problem texts paid off as new practical solutions to calculating Pythagorean triples and square roots. It is ironic that popular awareness of Babylonian mathematics is primarily because of the peculiar "choice" of base-60; geometric algebra was a much more notable achievement.

Although we now recognize that many developments previously credited to Greek mathematics and particularly to Pythagoras are of Babylonian origin, the successor Greek mathematics marks a watershed passage into maturity. In chapter 6, I shall briefly consider why, despite Babylon's newly recognized achievements, they will always remain overshadowed by Greek mathematics.

6

MATHEMATICS ATTAINS MATURITY: RIGOROUS PROOF

6.1 PYTHAGORAS

Pythagoras (580–500 BCE) is probably the most famous of all mathematicians because the *Pythagorean theorem* about right triangles, $a^2 + b^2 = c^2$, is usually the first math we learn with somebody's name attached to it. It also happens to be a very useful result. We have now seen that the Babylonians knew the Pythagorean theorem at least one thousand years before Pythagoras, although Pythagoras may have been the first to prove it. In addition, the greatest Babylonian mathematical contribution, geometric algebra, is traditionally known as *Pythagorean geometric algebra.* This misplaced credit is simply because the mathematics attributed to Pythagoras has been studied for some twenty-five hundred years. It is only in the last one hundred years that Babylonian mathematical tablets have been unearthed and translated.

About twenty-five hundred years ago, Pythagoras traveled to Babylon to learn the mystic teachings of the Magi (a continuing gift of the Magi is the word *magic*) and returned to Greece to found the mystic Pythagorean Brotherhood, whose motto was "All is number." There is no way of knowing whether Pythagoras represented the Babylonian mathematics he learned as his own or whether his disciples invented the legends crediting these discoveries to him. At least one story attributing the discovery of the Pythagorean theorem to him we know must be apocryphal. It relates that Pythagoras was so pleased by his discovery that he sacrificed a bull to the gods. Since the Pythagorean Brotherhood was vegetarian, in details at least, this story is false.

Legend also has it that Pythagoras was the first man to conjecture that the earth is a sphere. My guess is he knew this because he came from an island, Samos in the Aegean, from which he observed a circular horizon. Inland dwellers, unless they live by a very large lake, would not even be aware that there is a circular horizon. Pythagoras also saw boats sail out of sight over the horizon and then sail back into sight across the horizon. Thus, he knew that the horizon was surely not the edge of a flat-disk world. What he saw, which nobody before him had seen, was that a spherical earth explained these observations. What Pythagoras did not see was that his Pythagorean theorem would have allowed him to calculate the distance to the horizon and to calculate the circumference of the earth. Figure 6.1.1 shows the variables on which the distance to the horizon depends.

The observers' eyes are a height h above sea level; the distance to the horizon is d. The observer can also see the top of a mountain of height H above sea level that is a distance D over the horizon. The radius of the earth is R. The drawing is not to scale: $R >> $ all other variables.

FUN QUESTION 6.1.1: If you are standing on the shore so that your eyes are 2 m above sea level, how far away is the horizon? Use the result to devise a practical experiment to determine the diameter of the earth.

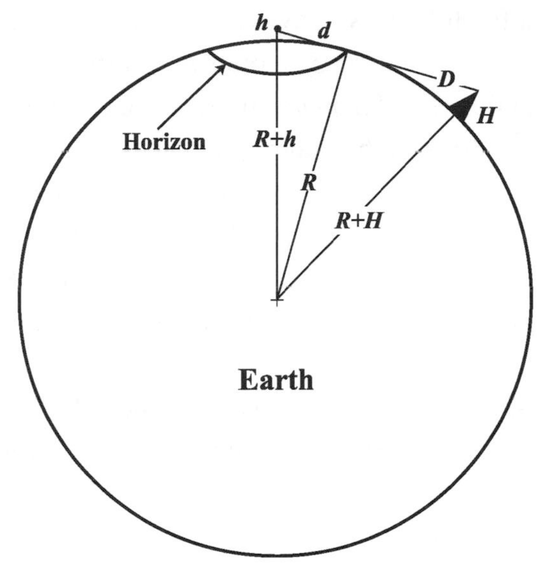

***Figure 6.1.1** Distance to the horizon*

FUN QUESTION 6.1.2: The ability of Stone Age Polynesian sailors to navigate to isolated islands in the Pacific Ocean was remarkable. One of the islands they discovered was Hawaii, on which stands Mount Mauna Kea, some 4,000 m above sea level. Due to the curvature of the earth's surface, what is the maximum distance from which Polynesian sailors could see the top of this mountain? Assume that from their sailing canoes their eye level was 2 m above sea level.

Members of the Pythagorean Brotherhood certainly did credit some of their own discoveries to their founder of godlike stature. In fact, crediting any mathematics to Pythagoras is just legend and tradition, with many contradictions. Ironically, this is because Euclid's *The Elements* (about 300 BCE) so completely summarized all prior Greek mathematics that most prior documents were rendered of historical importance only, and ceased to be copied. This is just like today's practice. As new and better texts become available, older texts go "out of print."

It is only in Euclid that we have documentation of the geometric algebra presented in chapter 5, which is traditionally credited to Pythagoras. Euclid's *The Elements* gives the diagram of figure 5.1.1 as Proposition 8, Book 2; the diagram of figure 5.1.2 as Proposition 4, Book 2; the diagram of figure 5.2.3 as Proposition 6, Book 2; and the Pythagorean theorem as Proposition 47, Book 1. We are thus assured that these constructions are of ancient origin and not inventions of more recent mathematicians; Euclid himself did not know their Babylonian origin.

Thales (624–548 BCE), possibly a teacher of Pythagoras, and Pythagoras are the first Greek mathematicians historically recognized. It has always been somewhat of a mystery how, with apparently no predecessors to build on, they could have made such profound progress. From the archeological discoveries of the last one hundred years, we now understand that they were expanding on a rich Babylonian inheritance. The extent of Greek inheritance from Egypt is still somewhat of a mystery, notwithstanding the opinion of Herodotus (see section 4.3).

Now that Pythagoras is no longer considered the discoverer of either the Pythagorean theorem or Pythagorean geometric algebra, does any significant mathematics remain to be credited to him? I will just consider one of the mathematical discoveries that legend credits to him, *figurate numbers*, in modern math jargon and illustrated in figures 6.1.2 and 6.1.3.

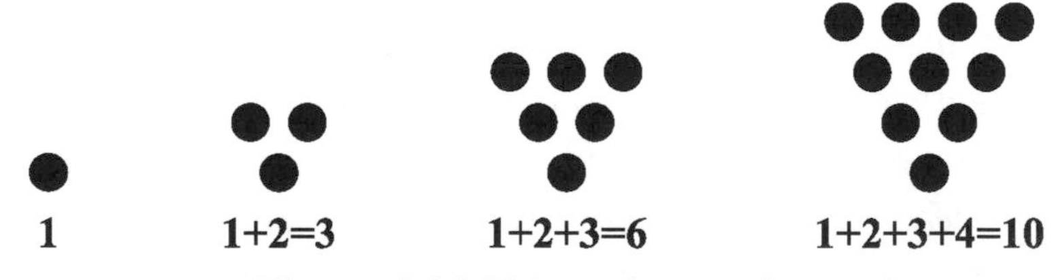

Figure 6.1.2 *Triangular numbers*

If n is the number of rows and S is the number of points in a triangular number, it is easy to see from the drawing that $S = 1 + 2 +$

$\ldots + n = n(n + 1)/2$, which is just the sum of an arithmetic series, previously derived algebraically as equation (1.3.2). Pythagoras also attributed mystical properties to triangular numbers. The four-row triangular number named *tetraktys* represented the four elements: fire, water, air, and earth. The three-row triangular number also had special significance as a *perfect* number: $1 + 2 + 3 = 1 \times 2 \times 3 = 6$. The mystic overlay to these results does not diminish that Pythagoras was developing a new, insightful way of thinking about numbers.

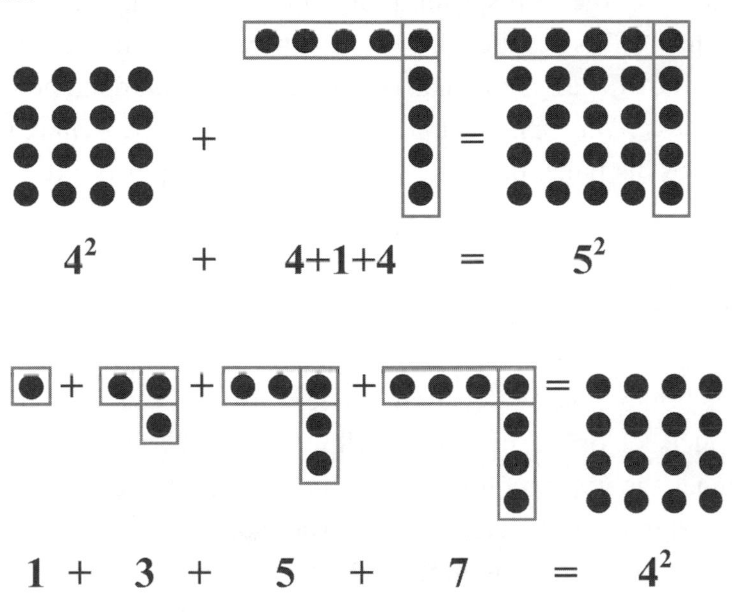

Figure 6.1.3 *Square numbers*

By adding a gnomon with $2n + 1$ points to a square of n rows, a square of $n + 1$ rows is formed and therefore: $n^2 + (2n + 1) = (n + 1)^2$. This result should look familiar because it is just the Babylonian, completing-the-square construction that we have seen so many times previously, as, for example, in figure 5.1.2 with $a = 1$ and $b = n$. But Pythagoras carried this construction further to obtain a completely new result: a series of n gnomons produces a series of n odd numbers such that: $S_o = 1 + 3 + 5 + \ldots + (2n - 1) = n^2$.

FUN QUESTION 6.1.3: Derive a general solution to the sum of the series of even numbers: $2 + 4 + \ldots + 2m$.

With these *figurate number* constructions, Pythagoras *digitalized* Babylonian geometric algebra and transformed it into *number theory*. The Babylonians did numerical calculations; Pythagoras and his successor Greeks did mathematics. Although Pythagoras himself may not have been the inventor of any truly great mathematics, his role in inspiring philosophers to think mathematically with his "All is number" creed was probably seminal to the development of Greek mathematics. Pythagoras also apparently coined the words *philosophy* and *mathematics*. Ironically, Pythagoras still deserves his fame, but not for discovering his eponymous theorem.

6.2 ERATOSTHENES

In section 5.2 we saw that all integers are either prime numbers or products of prime numbers. Prime numbers are fundamental to number theory, which of course is why the Greeks called them *prime* numbers. We also saw that Babylonian scribes usually only divided by *regular* numbers ($2^i3^j5^k$), which in the sexagesimal system are products of the prime factors of 60. A possible interpretation of this is that scribes were aware of the prime number concept, but the evidence is only that they at least came very close. Thus, even by about 2000 BCE the concept of prime numbers had probably not been discovered, justifying the questioning in section 2.1 of the conclusion by some that hunter-gatherers of twenty thousand years ago were aware of prime numbers. It is simply too sophisticated a concept for that stage of human evolution.

The first definitive evidence of the appreciation of prime

numbers is from the writings of the Greek mathematician Eratosthenes (250 BCE). He invented what is now called the *Sieve of Eratosthenes*, which is still a useful method for identifying prime numbers.

1	2	3	4	5	6	7	8	9	10
11	12	13	14	15	16	17	18	19	20
21	22	23	24	25	26	27	28	29	30
31	32	33	34	35	36	37	38	39	40
41	42	43	44	45	46	47	48	49	50
51	52	53	54	55	56	57	58	59	60
61	62	63	64	65	66	67	68	69	70
71	72	73	74	75	76	77	78	79	80
81	82	83	84	85	86	87	88	89	90
91	92	93	94	95	96	97	98	99	100

x2 x3 x5 x7

***Table 6.2.1** The Sieve of Eratosthenes for numbers up to 100*

Again, to illustrate the new insights and sophistication that differentiates Greek from Babylonian mathematics, table 6.2.1 presents the Sieve of Eratosthenes for numbers up to 100. To use the method of Eratosthenes, a table is composed of all numbers up to the highest number considered. The first prime number in the table is 2, and any multiple of 2 cannot be a prime number, so they are all eliminated from the table. The next noneliminated prime number is 3, and any multiple of 3 cannot be a prime number, so they are all eliminated from the table. This procedure is continued with 5 and 7 being identified as prime numbers, and their multiples are eliminated from the table. The numbers remaining are the prime numbers up to 100. It is not necessary to eliminate multiples of numbers greater than 7 because every number up to 100 that is not prime is eliminated as a multiple of a prime number that is not greater than 10 = $\sqrt{100}$ (see FUN QUESTION 3.2.3). Seven is the largest prime number less than 10.

FUN QUESTION 6.2.1: In order to construct a table of prime numbers up to 500 using the Sieve of Eratosthenes, multiples of what prime numbers must be eliminated?

The Sieve of Eratosthenes is easy to apply and to understand. Babylonian scribes could have invented it more than one thousand years earlier—but they apparently did not. Its invention was only possible after Pythagoras had made the study of properties of numbers a subject worthy of the attention of Greek philosophers.

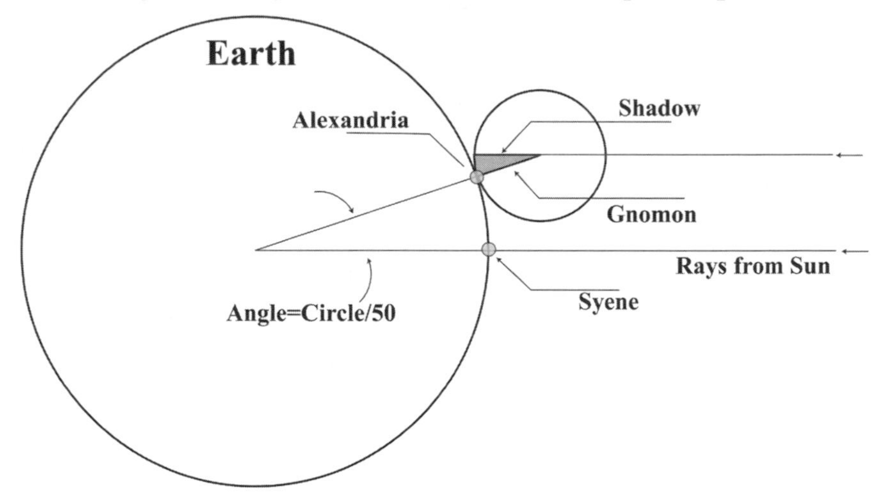

Figure 6.2.1 *Eratosthenes' measurement of the circumference of the earth*

The name Eratosthenes is not exactly a household word, although it is of course well known to students of mathematics, but it is also well known to every student who has taken a course in astronomy. Legend has it that although Pythagoras was the first to have conjectured that the earth was a sphere, Eratosthenes was the first to have measured its circumference. Figure 6.2.1 illustrates his method. The city of Syene in ancient Egypt, near the present-day city of Aswan, is located on the Tropic of Cancer. Thus, at noon on the day of the summer solstice (June 22, the longest day of the year) the sun is exactly overhead and a straight, vertical gnomon does not cast a shadow. A ray from the sun points along a radius of the earth, directly to the center of the earth.

On the day of the summer solstice (it does not even have to be in the same year), at high noon in Alexandria some 5,000 stadia to the north, a gnomon casts a shadow defining an angle of a fiftieth of a circle. By similar-triangle Euclidian geometry, this is just the angle defined at the center of the earth by radii extending to Syene and Alexandria. The circumference of the earth is therefore 50 × 5,000 = 250,000 *stadia*. One *stadion* (the length of an Olympic stadium) is equal to about 600 feet ≅ 180 m (see table 3.2.5), so that Eratosthenes determined the earth circumference to be 45,000 km, a very good approximation to the modern value of 40,000 km. At least some of the Greek philosophers were not solely engaged in abstract contemplation of mathematics, but were also able to put their newly found understanding of mathematics to practical use.

FUN QUESTION 6.2.2: At high noon on June 22 you measure an angle of 1/16 circle (360/16 = 22.5°) defined by the shadow of a vertical *gnomon*. You know it is June 22 because the shadow at high noon is the shortest on that particular day. You know when it is high noon because the shadow is shortest at that particular time of day. What is your latitude?

ANSWER: You are 22.5° north of the Tropic of Cancer, which in turn is 23.5° north of the equator; therefore, your latitude is 46° north of the equator.

The axis of rotation the earth tilts at is at an angle of 23.5° to the axis about which the earth circles around the sun. Thus only between the Tropic of Cancer (23.5° north latitude) and the Tropic of Capricorn (23.5° south latitude) can the sun ever be directly overhead. Eratosthenes did not understand that the earth circles the sun or that the axis of rotation of the earth was tilted. With the assumption that the earth was spherical, he was able to determine the circumference, but he could not have determined latitude, as done here.

The method of Eratosthenes, though clever, is not the simplest way that the circumference of the earth could have been measured at that time. Latitude is the angle between the direction to the North Star and to the northern horizon. By measuring latitudes at any two locations and taking the difference, and measuring the north-south distance between the two locations, it would have been possible to calculate the circumference on any clear night, and not just at high noon on the day of the summer solstice. You, too, can measure the circumference of the earth, without having to travel to Egypt, with a simple angle-measuring device (a protractor). Try it.

FUN QUESTION 6.2.3: At high noon on March 22 you measure an angle of 1/8 circle (360/8 = 45°) defined by the shadow of a vertical *gnomon*. What is your latitude?

Fortunately, in 1492 Christopher Columbus was not aware of Eratosthenes' calculation. His mistaken underestimate of the circumference of the earth allowed him an overly optimistic estimate of his chance for success in reaching China by sailing west, without which he possibly would never have attempted his voyage.

6.3 HIPPASUS

Hippasus (about 400 BCE) is also not a widely recognized name. Nevertheless, he is credited with an elegant proof that $\sqrt{2}$ cannot be a rational number, a concept of seminal importance in the development of mathematics. As shown in section 5.5, Babylonian scribes invented a clever algorithm to calculate $\sqrt{2}$ to surprising accuracy. This was certainly one of the high points of Babylonian mathematics. To help appreciate the greater depth and the subtlety of Greek thought, let us consider Hippasus's proof that $\sqrt{2}$ is an *irrational* number.

To prove that for the right triangle (a, a, c), $c/a = \sqrt{2} \neq p/q$, where p and q are integers, Hippasus used a method called *reductio ad absurdum*:

1. Assume that $c/a = p/q$ is a *rational, reduced fraction*, where p and q are integers with no common factors. Show that this leads to an absurdity.
2. With this assumption that $c/a = p/q$, $(c/a)^2 = (p/q)^2$.
3. For a triangle (a, a, c), the Pythagorean theorem yields $c^2 = 2a^2$ and therefore $p^2 = 2q^2$, and p^2 must be even because 2 is one of its factors.
4. Hence p must also be even and can be written as $p = 2m$, where m is some integer (even number × even number = even number; odd number × odd number = odd number).
5. Hence $p^2 = 4m^2 = 2q^2$ so that $q^2 = 2m^2$ and therefore q^2 and q must also be even.
6. In **Step 4** we proved that p must be even, but in **Step 1** we assumed that p/q was a *reduced fraction* and therefore ***q* must be odd** in order not to have a factor of 2 in common with p, but in **Step 5** we proved that ***q* must be even.** The assumption that c/a is a *rational fraction* must be wrong, and therefore $\sqrt{2}$ is an *irrational* number, QED.

Hippasus was a member of the Pythagorean Brotherhood. One legend has it that when Pythagorean colleagues learned of his proof, they drowned him for destroying their belief that everything could be explained by whole numbers and their ratios. The Pythagorean motto "All is number" was understood to mean, "All is *rational* number." Whether apocryphal or not, the story makes the point that some Greeks took their mathematics very seriously.

The logic in this proof is at a level of abstraction and difficulty that makes Babylonian mathematics look like child's play. Even more difficult to comprehend than the meticulous logic of this proof is the insight that motivated Hippasus even to consider that there was such an entity as an irrational number. It had been perceived that as many points as desired could be evenly spaced to divide the interval between 1 and 2, wherein $\sqrt{2}$ falls, so that it had been "obvious" that any number in the interval could eventually be defined by some rational-number point. Hippasus showed that the "obvious" dogma of the Pythagoreans, which they believed was the key to understanding the laws of nature, was wrong, and so they drowned him in the time-honored tradition of killing the messenger who brings bad news.

FUN QUESTION 6.3.1: Prove that even × even = even, and odd × odd = odd.

Now that we have examined both Egyptian/Babylonian and some Greek mathematics, let us revisit one of those academic squabbles that make history more interesting. In 1962 Professor of Mathematics at New York University Morris Kline published his *Mathematics: A Cultural Approach*, in which he wrote, "The mathematics of the Egyptians and the Babylonians, is the scrawling of children just learning to write, as opposed to great literature." The evidence discussed in this book, which has the advantage of more recent Egyptian/Babylonian discoveries, definitely supports Kline's view.

However, some authors have taken umbrage with Kline's supposed slight to Egyptian/Babylonian accomplishments. Richard Gillings's *Mathematics in the Time of the Pharaohs* (1972) and G. G. Joseph's *The Crest of the Peacock: Non-European Roots of Mathematics* (1991), both of which I have found valuable and have cited frequently in the notes and references section, were critical of Kline's opinion.

There is a tendency for every scholar to overrate the importance of his particular specialty. Gillings, quite understandably, took personally a supposed aspersion on ancient Egyptian accomplishments. Joseph believed that Kline exhibited a generally prevalent "Eurocentric" prejudice. Kline's statement may not have been politically correct, but Greek mathematics clearly does represent a quantum jump beyond any predecessor mathematics. The important question is not whether credit was justly apportioned, but why Greek mathematics was so remarkable. What was different in Greece, or about the Greek people?

7
WE LEARN HISTORY TO BE ABLE TO REPEAT IT

7.1 Teaching Mathematics in Ancient Greece and How We Should but Do Not

Rather than the title of this chapter, the usual pithy quotation about history is that we learn it in order *not* to repeat it. This implies that what is important is to learn the blunders, and being aware of them, they can be avoided. Alas, blunders are in such a large part due to unavoidable stupidity and compulsions that, aware of them or not, we keep on repeating them.

It would be nice to understand what made possible the transition from the largely intuitive arithmetic of the Egyptians and Babylonians to the concept of rigorous proof and the elegant mathematics of the Greeks.

Simple passage of time cannot explain the Greek transition because mathematics stagnated in Egypt and Babylon in the millennium or so that separates Greek from Egyptian and Babylonian inventions. Nor can it be explained as ethnic: Athens epitomized Greek intellectuality and grandeur, but life was Spartan only about 100 km away. The Greek miracle, as it has been called, was not limited to mathematics; it included literature, philosophy, history, politics, art, science, and mathematics. It was due to a confluence of historical and societal factors.

Alphabetic writing was probably critical. A Phoenician invention from around 1000 BCE, it made reading/writing much more efficient. The Greeks, with the improvement of adding vowels to the consonants-only Phoenician alphabet, also soon adopted alphabetic writing. Fortuitously, in the ninth or eighth century BCE the poet Homer canonized Greek legends and mythology in his epics, the *Iliad* and the *Odyssey*. These works served to standardize the Greek language, although they were probably not written down until the seventh century BCE. More important, they became a kind of Bible for Greek society. The *Iliad* and the *Odyssey* taught the essential lessons of Greek pride, honor, good behavior, and morality, but they were not sacerdotal. Greek thinking was not overly burdened by priestly orthodoxy.

In Greek society, literacy was not limited to a small class of scribes who mostly performed bureaucratic trivia, as had been the case in Egypt and Babylon. In Greece, slaves performed much of the menial labor, which made possible the existence of a literate, leisure class. Into this intellectually receptive class, Pythagoras, probably more than anyone else, made mathematics a subject worthy of study. Somewhat ironically, his simplistic but appealing concept that understanding numbers would provide the answer to the mysteries of existence performed a great service to Greek science. Pythagoras attracted the attention of the scientifically curious to productive mathematics rather than to unproductive physics.

Today we turn to physics for basic answers (the intellectually lazy or weak turn to mysticism), but physics without experimentation to confirm or falsify its theories easily strays from reality. The Greeks had very few tools of experimental physics, but as illustrated previously here by Pythagoras's conjecture that the earth is a sphere and Eratosthenes' determination of the circumference of the earth, they used the few tools that were available.

It was not until the some two thousand years after the "golden age of Greece" that technology could provide the instrumentation that enabled Galileo (1564–1642) and Newton (1642–1727) and their successors to build a solid base of experimentally tested theories. Untested theories of the unquestionably brilliant Aristotle (484–322 BCE) misled scientists for many centuries until falsified by Galileo and Newton.

It is interesting to compare Greek history with that of a neighboring culture which evolved with an amazingly similar timeline, that of the Jews in Canaan. Hebrew, the language of the Jews, also adopted alphabetic writing from the Phoenicians around 1000 BCE. The history and the beliefs of the Jews were canonized with the anonymous writing of Deuteronomy (Old Testament), or Devarim (Hebrew Bible). This was first written down in the seventh century BCE, the same century in which Homer's canonization of Greek legends and myths was first written down. The differences in these two canons determined entirely different intellectual directions for Greeks and Jews. While the Greeks were free to inquire wherever their curiosity led them, the Jews devoted themselves to the solution of just one problem: What is the meaning of the fine print in their covenant with God? So completely were the Jews uninvolved with the mathematics that one of the very few contributions they made was to explain why some people gave the value of π as 3, while others gave a value of 3 1/8: "Obviously," said the Rabbi, "some people measure the circumference inside the circle, while others measure the circumference outside the circle."

An interesting contrast between ancient Greek and Jewish cultures can be made when near the end of the sixth century BCE both Pythagoras and Jewish exiles (see section 3.2) were in Babylon. As we have seen, Pythagoras returned from Babylon with much of the mathematics now credited to him. The Jews returned with new interpretations of their covenant with God, epitomized in the memorable verses:

> By the waters of Babylon, there we sat down and wept when we remembered Zion.
> On the willows there we hung our lyres.
> For there our captors required of us songs.
> And our tormentors, mirth, saying, "Sing us one of the songs of Zion!"
> How shall we sing Yahweh's song in a foreign land?
> If I forget you, O Jerusalem, let my right hand wither!
> Let my tongue cleave to the roof of my mouth . . .
>
> Psalms (Old Testament); Tihilim (Hebrew Bible)

In ancient Greece, mathematics was a respected subject. Now it is the most detested of school subjects. What did the Greeks do in mathematical education that we are not doing today? To describe what we are doing today, I quote from the book, *How to Solve It* (1945), by the noted mathematician G. Polya. "[The teacher] fills his allotted time with drilling students in routine operations that kills their interest, hampers their intellectual development. . . . Future teachers pass through elementary schools learning to detest mathematics. . . . They return to elementary school to teach a new generation to detest it."

Greek schooling was limited to reading, writing, rhetoric, music, and athletics. After completion of schooling, a man could add to his education by listening to lectures given by wandering scholars called Sophists. If he specifically wanted to learn mathematics, he listened to a Sophist who lectured about mathematics. Mathematics was an elective for the mature. Arithmetic ability was not necessary for the educated Greek; he had slaves to do his menial mental as well as physical labor. Today's more humane equivalent of the slave is the electronic calculator.

If we go back in history only some few hundred years to when all countries had agrarian economies, general arithmetic competence was low. Only following the industrial revolution was there a need for more general arithmetic competence, and we entered our present cycle of intensive instruction of children. When pencil/paper was the only method of doing arithmetic, there was some logic to this. With the invention of the electronic calculator, and its miniaturization and ubiquitous presence, the justification for requiring perfect memorization of multiplication tables and extensive drilling in arithmetic has long passed.

Emphasis on a facile pencil/paper arithmetic ability that relies on memorization not only teaches children to detest mathematics, but it also teaches them to use the wrong part of their brains when they eventually try to learn algebra, geometry, and other mathematics. Ask just about any high school graduate, who had successfully completed a course in algebra more than a few years previously, how to solve the simple quadratic equation, $x^2 + x = 1$? (See FUN QUESTION 5.4.2). The typical answer is, "I remember that it was solved by an equation like, something plus or minus the square root of something. . . ." Invariably, they were trying to do mathematics the same way they had learned to do arithmetic, by memorization. Their high school algebra course had been a total waste of time, although they had remembered "the equation" long enough to pass final exams with sufficient grades for the school administration to be able to point with pride at the quality of the mathematics program.

The process of eliminating some of the obviously obsolete, pencil/paper arithmetic has begun. When I was a student, probably in about sixth grade, we still learned how to do pencil/paper square roots. It was by a method that somewhat resembled long division.

We learned the method completely by rote, without any pretense of understanding.

Few of the children who learned this square root calculation ever used it again. Like anything learned strictly by memory, without understanding its logic and then seldom used, it was quickly forgotten. Why was it taught?

FUN QUESTION 7.1.1: Derive a pencil/paper, greedy-algorithm calculation for the square root of a six-digit number, $N = a_5a_4a_3a_2a_1a_0$, to three-digit precision. Let $(A_2A_1A_0)^2 \cong N$, so $A_2A_1A_0 \cong \sqrt{N}$. Use equation (1.3.1): $(A_2A_1A_0)^2 = (10^2A_2 + 10A_1 + A_0)^2 \cong N$. Test the method by calculating the square root of the numbers 123,456, 12,345.6, and 1,234.56.

Nowadays, we do square roots by simply pressing the key on a calculator with the $\sqrt{\ }$ icon, and the idiocy of teaching this obsolete, by-rote, pencil/paper calculation is sufficiently obvious that it has been largely eliminated from present-day instruction. However, present-day instruction in arithmetic with common fractions, frequently a difficult and unpleasant learning experience, is less obvious but equally obsolete arithmetic. Consider the way we learn to evaluate a common-fraction expression such as (5/18 – 5/99). First, we learn to factor and separate out common factors:

$$5/18 - 5/99 = 5/9(1/2 - 1/11)$$

Then we learn to convert fractions to common denominators:

$$(5/9)(1/2 - 1/11) = (5/9)(11/22 - 2/22) = (5/9)(9/22) = 5/22$$

But even 5/22 is probably not a very useful answer, so we then calculate its decimal value (most likely using an electronic calculator) as $5/22 = 0.2\underline{27}$, which, depending on the precision required, we may express approximately as 2/10 or 23/100 = 23 percent, or as 227/1,000 (the way baseball batting averages are expressed). So why not simply just learn to do the entire calculation using an electronic calculator, just keying in a sequence of numbers and operations: $5 \div 18 - 5 \div 99 = 0.2\underline{27}$? With the ubiquitous availability of electronic calculators, arithmetic with common fractions is just as obsolete as calculating square roots.

The operations of factoring and common denominators are no longer required for basic arithmetic, but they are necessary in algebra where we do not have the luxury of being able to enter numbers into an electronic calculator. Consider, for example, the algebraic expression

$$\frac{1}{a+b} + \frac{2b}{a^2 - b^2}$$

To simplify the expression, first factor and obtain

$$\frac{1}{a+b} + \frac{2b}{(a+b)(a-b)}$$

Next, separate out the common term and obtain

$$\frac{1}{a+b}\left(1 + \frac{2b}{a-b}\right)$$

Finally, write the expression in parentheses with a common denominator and obtain

$$\frac{1}{a-b}$$

as the result of the simplification. To obtain this useful result, we used exactly the same operations used just previously to do common-fraction arithmetic. Why not defer learning these operations until algebra, when students will appreciate the importance of these operations rather than perceiving them as ridiculous, obsolete arithmetic?

With the elimination of recognized, obsolete arithmetic, will the time devoted to arithmetic instruction be decreased? Of course not. The problem is not just the identification and elimination of obsolete arithmetic. The problem is also to restrain the educational bureaucracy from filling in the saved instruction time with other mathematical material that is just as useless, boring, and detested.

Another problem is that many parents do not know the difference between arithmetic and mathematics, or the difference between memorizing and understanding. Ironically, they will consider their child's inability to reel off the multiplication table and similar arithmetic feats as a defective education, even though they themselves now use an electronic calculator to multiply any product greater than 2×3.

Learning to read and write is easy and enjoyable for most children. This is because children start school with knowledge of language that has been intuitively acquired, and they are eager to learn how to extent this obviously very useful skill. If we would wait a few years before beginning so much formal instruction in arithmetic, we would similarly have children starting with intuitively acquired knowledge about numbers and arithmetic and eager to learn how to extend this obviously useful skill.

The instinctive answer of any bureaucracy to a problem is always *more,* but mathematics education would be better served by replacing with strenuous athletic play the time devoted to pencil/paper arithmetic that is no longer required in our electronic calculator era.

Mathematical progress clearly should not be measured by the mathematical ability of the general public; mathematical progress is determined by the efforts of the relatively few mathematically gifted. Presently, and even more in the future, the inventions of calculators and computer programs by a mathematical elite relieve the rest of us of a need for much mathematical competence.

A counterproductive bureaucratic solution for improving mathematics is frequent standardized nationwide tests. Every student soon learns the how-to-pass-tests algorithm: *When understanding fails, memorize. When memorization fails, cheat.* These tests measure teacher success, and their very jobs depend on their class's performance; they also know the how-to-pass-tests algorithm. Math teachers know that most students will get better grades on these tests if the teacher emphasizes memorization rather than understanding. They cheat by teaching how to answer the type of question that they know will appear in such tests. They cheat by recommending that their poorer students stay home on test day.

Certainly, a system cannot be fixed without testing it, but this does not necessarily require more standardized testing. You can measure how good a specific teacher is by how many of the students choose subsequent math electives. You can measure the overall quality of a mathematics program by how many students choose math or physics in college. Students choose courses they enjoy; nobody enjoys math if they do not understand what they are doing. Understanding is what should be tested, not ability to pass a standardized test. The solution is not more standardized testing, but more math electives for the interested and gifted.

However, it cannot hurt to make mathematics more interesting. I hope that this book provides mathematical history and insights that can do just that.

Ω

APPENDIX: ANSWERS TO FUN QUESTIONS

1.1.2 (p. 24): Generalizing algebraically, a square of sides $\sqrt{ab}$ has the same area as a rectangle of sides *a* and *b*. Probably the only way you know how to calculate the $\sqrt{8}$ is by using the key with the $\sqrt{\ }$ icon on an electronic calculator. Only two generations ago calculation of square roots was taught in elementary school, another example of now-obsolete arithmetic. Because each generation has little knowledge of what was taught in previous generations, we tend not to perceive the ongoing evolution in something so apparently fundamental as arithmetic.

1.2.1 (p. 28): There are ten ways of exhibiting the number nine:

01111 11111 10111 11111 11011 11111 11101 11111 11110 11111
11111 01111 11111 10111 11111 11011 11111 11101 11111 11110

There are forty-five ways of exhibiting the number two. There is an elegant combinatorial-mathematics solution to this problem, but the simple yet tedious method given below also works.

With a 1 in the first position on the right, there are nine possible positions for the other 1:

00000 00011, 00000 00101, and so on. With a 1 in the second position on the right, there are eight possible positions for the other 1: 00000 00110, 00000 01010, and so on. Continuing with this process, the total number of arrangements is 9 + 8 + 7 + 6 + 5 + 4 + 3 + 2 + 1 = 45.

1.2.2 (p. 30): For each exhibition of the left hand, there are five ways of exhibiting 4 on the right hand: 01111, 10111, 11011, 11101, and 11110. Since there are the same five ways of exhibiting 20 on the left hand, the total number of exhibitions is 5 × 5 = 25.

1.2.3 (p. 30): A left hand of 80 + 40 + 20 + 10 + 5 maximizes the counting limit.

1.2.4 (p. 30):

Count total	Finger display
1	00000 0000**1**
2	00000 000**1**0
3	00000 00011
4	00000 00**1**00
5	00000 00101
6	00000 00110
7	00000 00111
8	00000 0**1**000
...	...

Table FQ Answer 1.2.4a

Count total	Finger display
16	00000 **1**0000
32	0000**1** 00000
64	000**1**0 00000
128	00**1**00 00000
256	0**1**000 00000
512	**1**0000 00000

Table FQ Answer 1.2.4b

The pattern of finger values for nonredundant counting is already clear. The value of each finger is twice the value of the preceding finger. Now it is possible to assign values to the all of the remaining fingers.

It is also easy to see the pattern; the value for each finger equals the sum of the values of all preceding fingers plus one. Since the value of the tenth finger (10000 00000) is 512, then the sum of all preceding fingers (01111 11111) is 511, so ten fingers can count to 512 + 511 = 1,023. The replacement number for this counting is 1-for-2. No other finger counting that exhibits a sequence of extended and nonextended fingers can count higher because for every number there is one and only one possible finger arrangement.

1.3.2 (p. 40): Two pairs of hands. One way of solving this is by using the results from FUN QUESTIONS 1.2.4 and 1.3.1: The number of different arrangements of one pair of hands is 1,024. By the multiplication rule, the number of arrangements of two pairs of hands is $1{,}024 \times 1{,}024 = 1{,}048{,}576$. The highest number that can be counted is one less than this.

1.3.4 (p. 47): The base-2 addition table is simply one entry:

+	1
1	10

Table FQ Answer 1.3.4

$$\begin{array}{r} 11010100 \\ +\ 10101111 \\ \hline 110000011 \end{array}$$

$$110000011 = 603_8 = (6 \times 8^2) + 3 = 387_{10}$$

$$110000011 = 183_{16} = 16^2 + (8 \times 16) + 3 = 387_{10}$$

While it is also possible to convert directly from base-2 to base-10, it is much easier to convert from base-2 to base-8 or base-16. Conversion from base-2 to base-8 or base-16 is simply by inspection while conversion from base-2 to base-10 requires calculation.

1.3.5 (p. 47): The base-8 addition table is:

+	1	2	3	4	5	6	7
1	2	3	4	5	6	7	10
2		4	5	6	7	10	11
3			6	7	10	11	12
4				10	11	12	13
5					12	13	14
6						14	15
7							16

Table FQ Answer 1.3.5

$$\begin{array}{l} 234_8 \\ +\ \underline{456_8} \\ 712_8 = 111001010_2 = (7 \times 8^2) + 8 + 2 = 458_{10} \\ 111001010_2 = 1CA_{16} = 16^2 + (12 \times 16) + 10 = 458_{10} \end{array}$$

1.3.6 (p. 47): Is there an integer, b, such that $2b^3 + 4b^2 + 3b + 1 = 366_{10}$? Apply the greedy algorithm: The greediest possible value for b^3 must be less than 366/2 = 183. The largest integer whose cube is less than 183 is 5; $5^3 = 125$. Check to see whether this greediest choice is the correct solution: $2{,}431_5 = (2 \times 5^3) + (4 \times 5^2) + (3 \times 5) + 1 = 366_{10}$. At least from this threat we are safe.

2.2.1 (p. 55): Right knee + left elbow = man + right thumb. The *replacement*, 1 man = 33, now makes this a *positional* system that is capable of counting to higher than 33.

2.2.2 (p. 58): Table FQ Answer 2.2.2 defines a base-3 system that conforms to equation (1.3.1). Note that it was necessary to define a new word, threethree, for 3^2, just as in base-10 a new word, hundred, is defined for 102. Perhaps this analogy will be more obvious if instead of using the word hundred, the word tenten is used.

Base-10 Number	**Aborigine Language**	**Base-3 Interpretation**
1	one	1
2	two	2
3	three	3
4	three one	$3 + 1$
5	three two	$3 + 2$
6	two three	2×3
7	two three one	$2 \times 3 + 1$
8	two three two	$2 \times 3 + 2$
9	threethree	9
10	threethree one	$9 + 1$

Table FQ Answer 2.2.2

2.2.3 (p. 58):

Base-10 notation	Base-3 notation
1	1
2	2
3	10
4	11
5	12
6	20
7	21
8	22
9	100
10	101

Table FQ Answer 2.2.3

3.1.1 (p. 70):

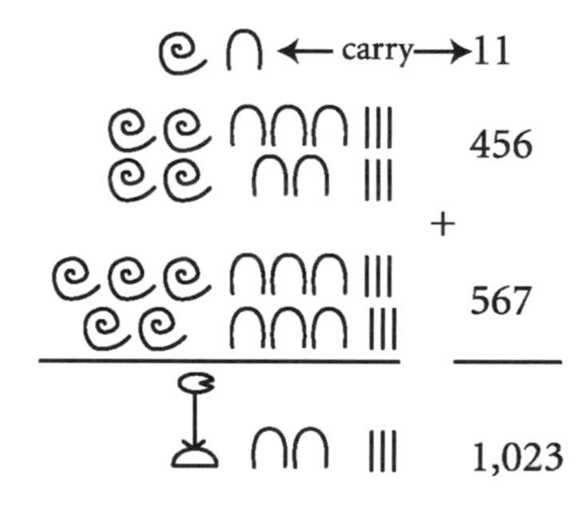

Figure FQ Answer 3.1.1

Despite the fact that hieroglyphics is *additive*, addition is carried out essentially as we do with our *positional* base-10 system. The general equation (1.3.1) that describes any positional system, describes additive number systems just as well. Only a slight modification of the meaning of the terms is required. For an additive number system the general term a_i is now simply the number of symbols of value b^i, rather than the digit in position i.

3.1.2 (p. 70):

carry
10 100
6 50 400
7 60 500
3 20 1,000

Figure FQ Answer 3.1.2

Note that while hieroglyphics was written from either left to right or from right to left, hieratics was always written from right to left, just as in other Semitic languages such as Hebrew and Arabic. Hieratic numbers thus appear to us to be written backward. Since hieratic numbers are additive, no zero symbol is required.

3.2.1 (p. 88):

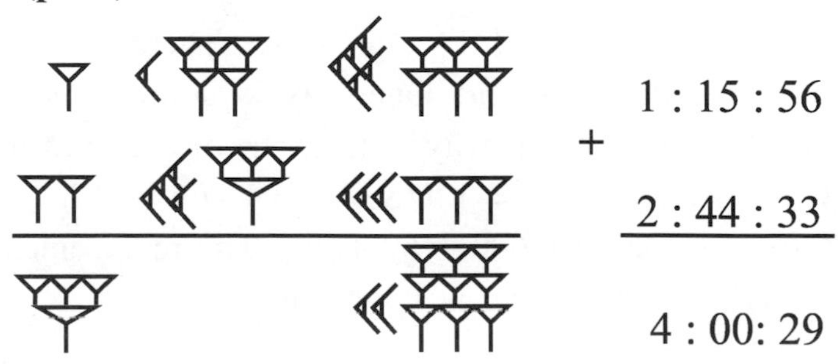

Figure FQ Answer 3.2.1

3.2.2 (p. 90): Finding all the divisors might at first glance appear to be a tedious process. However, divisors always come in pairs: (1, 240), (2, 120), and so on. The smaller divisor of a pair will equal the larger divisor of a pair when they both equal $\sqrt{240}$. Clearly, the smaller divisor of a pair cannot exceed $\sqrt{240}$, and the larger divisor of a pair cannot be less than $\sqrt{240}$. Since $\sqrt{240} = 15.4$, only lower divisors less than 16 need be tested, not so tedious. Thus, one readily finds nine divisor pairs: (1, 240), (2, 120), (3, 80), (4, 60), (5, 48), (6, 30), (8, 30), (10, 24), (12, 20).

3.2.4 (p. 96): One proof is to let $m = M + n$, and use the answer to FUN QUESTION 3.2.4 so that: $b^m/b^n = b^{M+n}/b^n = b^M\ (b^n/b^n) = b^M = b^{m-n}$, QED.

3.2.5 (p. 96): One proof is to use the answer to FUN QUESTION 3.2.4 so that: $b^m = b^{m/2} \times b^{m/2}$. Take the square root of both sides and $\sqrt{b^m} = b^{m/2}$, QED. Thus, another way of writing the square root of a number b is as $b^{1/2}$. Power notation is not only a convenient way of writing numbers; it also provides convenient arithmetic.

3.2.6 (p. 99): Base-30.

3.2.7 (p. 107): Use the greedy algorithm. 874 *nindan* = 14 *USH*, 3 *eshe*, 4 *nindan*.

3.2.8 (p. 112): 123 horses horses. There are two possible ways of writing 43 *iku*: One way is simply 43 *iku* = *iku*. Another way is in terms of standard mixed units: 43 *iku* = 4,300 *sar* = 1 *USH* + 1 *eshe* + 100 *sar* = (3,600 + 600 + 60 + 40). The replacement sequence for barley seed volumes is 1-for-5 and 1-for-6.

3.3.1 (p. 126): 780 = 2:3:0 = = 2 years + 60 days.

3.3.2 (p. 127): 1 min = 60 s, 1 hr = 60^2 s, 1 day = 24×60^2 sec, 1 year = $365 \times 24 \times 60^2$ = 31,536,000 s. If the vision of the metric system purists of the French Revolution had been accepted, we would now have that 1 min = 100 s, 1 hr = 100 min, 1 day = 10 hr and therefore 1 year = 36,500,000 s.

4.1.1 (p. 137): Using the greedy algorithm, $137 = 128 + 8 + 1 = 2^7 + 2^3 + 1$. It takes 7 additions to multiply by 128, and then 2 more additions to add multiplications by 8 and 1 to it, for a total of only 9 additions. If calculated by the modern multiplication method, 137×137 requires 9 multiplication operations plus 2 addition operations for a total of 11 operations. Despite this apparent advantage of Egyptian multiplication, the modern method of memorizing a multiplication table is generally much more efficient because application of the greedy algorithm is somewhat of a trial-and-error procedure and is therefore slow.

4.2.1 (p. 143): $3/7 = 1/7 + 2/7 = 1/4 + 1/7 + 1/28$, $4/7 = 2 \times 2/7 = 2(1/4 + 1/28) = 1/2 + 1/14$, $4/5 = 2 \times 2/5 = 2(1/3 + 1/15) = 2/3 + 2/15 = 2/3 + 1/10 + 1/30$.

4.2.2 (p. 144): Let $p/q = 0.\underline{074}$, then $1{,}000p/q = 74.\underline{074}$. Thus $1{,}000p/q - p/q = 999p/q = 74$, and so $p/q = 74/999$. Although 74/999 is a correct answer, what is usually wanted is the common fraction in its *reduced form* with all common divisors of numerator and denominator canceled out. To do this both numerator and denominator must be *factored*, which can sometimes be a difficult operation. In the present case factoring yields: $74/999 = (2 \times 37)/(9 \times 111) = (2 \times 37)/(9 \times 3 \times 37) = 2/27$. Note that only odd denominators occur in the table of $2/n$. This shows that the Egyptians understood the *reduced fraction* concept. It is not a profound mathematical concept, but it shows that they understood something about properties of numbers.

4.2.3 (p. 149): 5 11/16 = 5 + 1/2 + 1/8 + 1/16 = 101.1011_2. Note that the division of inch markings on a ruler provides an easy, graphic way of converting from a proper fraction to binary unit fractions.

4.2.4 (p. 149): 56.5 and 1011.011_2 = 11 + 1/4 + 1/8 = 11 3/8.

4.2.5 (p. 150): 0.0100001_2. Like fractions in the decimal system, or fractions in any number system with a base, the first nonzero digit to the right of the separation point unambiguously and simply identifies the larger fraction.

4.2.7 (p. 156): 2/5 = 1/3 + 1/15; 2/9 = 1/5 + 1/45 (greedy) = 1/6 + 1/18 (best).

4.2.8 (p. 156): 1/16 + 1/128 + 1/512 (binary); 1/5 + 1/45 (greedy); 1/6 + 1/18 (best).

4.2.9 (p. 157): Cut each melon into three pieces (halving each melon would only produce 38 pieces so that each child would not get a piece): each child gets a 1/3 piece; the remainder is 17 pieces of 1/3 melon ($[3 \times 19] - 40 = 17$). Cut each remainder piece into 3 pieces: each child gets a 1/9 piece; the remainder is 11 pieces of 1/9 melon ($[3 \times 17] - 40 = 11$). Cut each remainder piece into 4 pieces: each child gets a 1/36 piece; the remainder is 4 pieces of 1/36 melon ($[4 \times 11] - 40 = 4$). At this point, the division would probably stop because by now the remainder is negligible. But in the nonreal world of mathematics, the division can continue. Cut each remainder piece into 10 pieces and give each child a 1/360 piece. Thus 19/40 = 1/3 + 1/9 + 1/36 + 1/360. Although similar, this method is not the same as using the greedy algorithm and sometimes gives different results.

4.3.1 (p. 162): The largest number is 59:59:59. Thus $a = 59$, $r = 60$, $N = 3$, and $S = 60^3 - 1$, just as we had intuitively deduced in FUN QUESTION 1.3.1.

4.3.2 (p. 162): $a = 1/2$, $r = 1/2$, and therefore equation (4.3.1) yields $S = 1 - 2^{-N}$. Clearly, as $N \to \infty$ (read: as N approaches infinity) $2^{-N} \to 0$.

4.3.3 (p. 164):

1. $2N/3$

2. $N + 2N/3 = 5N/3$

3. $(5N/3)/3 = 5N/9$

4. $5N/3 - 5N/9 = 10N/9 = A$

5. $A - A/10 = 10N/9 - (10N/9)/10 = 10N/9 - N/9 = N$.

This general proof is easy because I used *algebraic notation.* Some Egyptian scibe composed this trick about thirty-five hundred years before the invention of algebraic notation, and he only had very awkward ways of expressing numbers and doing arithmetic. Composing this trick thus shows some insight into how numbers behave and is not as trivial as it might appear at first.

4.3.5 (p. 168): Babylonian V(frustum) $= h(a^2 + b^2)/2 = 6(2^2 + 4^2)/2 = 60$; Egyptian V(frustum) $= 56$. Percent error $= 100 \times 4/56 \cong 7\%$. Not a bad approximation at all, and easier to calculate.

4.3.8 (p. 174): 88.85 and 91.17 degrees, not particularly good approximations. However, the average angle is (88.85 + 91.17)/2 = 90.01 degrees, which is an excellent approximation.

4.4.1 (p. 180): 2 palms + 2 fingers = 10 fingers. 10/28 = 5/14 = 1/3 + 1/42 (by greedy algorithm).

5.1.1 (p. 190):

Decimal	Decimally transcribed sexagesimal	Cuneiform sexagesimal	Binary addition
173	2:53	vv <<< vvv <<	1x
+173	+2:53	+ vv <<< vvv <<	
346	5:46	vvv <<<< vvvv vv vv	2x = 1x + 1x
+346	+5:46	+vvv <<<< vvvv vv vv	
692 +692	11:32 +11:32	< v <<< vv +< v <<< vv	4x = 2x + 2x
1,384 +692	23:04 +11:32	<< vvv vvvv + < v <<< vv	8x = 4x + 4x
2,076	34:36	<<< vvvv <<< vvvv vv	12x = 8x + 4x
+173	+2:53	+vv <<< vvv <<	
2,249	37:29	<<< vvvv << vvv vvv vvv vvv	13x = 12x + 1x

Table FQ Answer 5.1.1

5.1.2 (p. 191): If there were a 57 × table, then 57 × 10 could be read directly from it. But 57 is not a regular number, and so there is no such table. The answer can be obtained from a 10× table:

10 × 57 = (10 × 7) + (10 × 50) = 70 + 500 = 1:10 + 8:20 = v < + vvvv <<
vvvv

570 = 9:30 = vvvv <<<
vvvv
v

5.1.3 (p. 195):

52			<<<< <	vv
37			<<<	vvvv vvv
$7 \times 2 = 14$			<	vvvv
$7 \times 50 = 350$		vvvv v	<<<< <	
$30 \times 2 = 60$		v		
$30 \times 50 = 1{,}500$	<<	vvvv v		
	<<<	vv		vvvv
$14 + 350 + 60 + 1{,}500 = 1{,}924 =$	3×600	$+ 2 \times 60$	$+ 0 \times 10$	$+ 4$

Table FQ Answer 5.1.3

5.1.4 (p. 198): $17{:}59 \times 12{:}46 = (59 \times 46) + ([17 \times 46] + [59 \times 12])1{:}0 + (17 \times 12)1{:}0{:}0$. Now use equation (5.1.1) to evaluate the four products:

$$59 \times 46 = (1{:}45^2 - 13^2)/4;\ 17 \times 46 = (1{:}03^2 - 29^2)/4;\ 59 \times 12 = (1{:}11^2 - 47^2)/4;\ \text{and}\ 17 \times 12 = (29^2 - 5^2)/4$$

Assuming that tables of squares are available up to $1{:}58^2 = 118^2$ and converting to decimal notation so that an electronic calculator can be used to calculate squares:

$$59 \times 46 = (105^2 - 13^2)/4 = 2{,}714 = 45{:}14;\ 17 \times 46 = (63^2 - 29^2)/4 = 782 = 13{:}02;\ 59 \times 12 = (71^2 - 47^2)/4 = 708 = 11{:}48;\ 17 \times 12 = (29^2 - 5^2)/4 = 204 = 3{:}24$$

The required product is:

$$45{:}14 + 13{:}02{:}0 + 11{:}48{:}0 + 3{:}24{:}0{:}0 = 3{:}49{:}35{:}14 = 826{,}514_{10}$$

The conversion of decimal numbers to decimally transcribed sexagesimal is made as usual using the greedy algorithm. I have glossed over the division by 4 required in this table-of-squares method of multiplication. However, division was not a trivial operation for the Babylonians and is considered in detail in section 5.2.

Assuming that tables of squares only up to 59^2 are available, then every number greater than 1:0 = 60 must also be squared using equation (5.1.1), but first it is necessary to convert every term $(a + b)^2 = (1{:}n)^2$ that is greater than 60^2 into $n^2 + 2{:}0n + 1{:}0{:}0$, which now has no squares greater than 59^2:

$$1{:}45^2 = 45^2 + 1{:}30{:}0 + 1{:}0{:}0 = 33{:}45 + 1{:}30{:}0 + 1{:}0{:}0 = 3{:}3{:}35 = 11{,}025$$
$$1{:}03^2 = 3^2 + 6{:}0 + 1{:}0{:}0 = 1{:}6{:}9 = 3{,}969$$
$$1{:}11^2 = 11^2 + 22{:}0 + 1{:}0{:}0 = 2{:}1 + 22{:}0 + 1{:}0{:}0 = 1{:}24{:}1 = 5{,}041$$

Clearly, much less effort is required if tables of squares up to 118^2 are available.

5.1.5 (p. 199):

- For $b + a$ to be odd, one variable must be even and the other odd. Let $a = 2m$ be even and $b = 2n + 1$ be odd, where m and n are integers. Thus $b + a = 2(n + m) + 1$ and $b - a = 2(n - m) + 1$, and hence both are odd.
- $ab = 2m(2n + 1) = 2m \times 2n + 2m = a(b - 1) + a$. With only the addition of a few easy additions to the multiplication algorithm, it is possible to use equation (5.1.1a) even for cases when $a + b$ is odd, and with only tables of squares of integers up to 59.

5.2.1 (p. 212): Let $R_1 = 2^i3^j5^k$ and $R_2 = 2^l3^m5^n$. Using the rule for multiplying powers (see FUN QUESTION 3.2.4), $(2^i3^j5^k)(2^l3^m5^n) = 2^I3^J5^K$, a regular number with $I = i + l$, $K = j + m$, and $L = k + n$, QED.

5.2.2 (p. 214): **1.** Use greedy algorithm: $125 = 2 \times 60 + 5 = 2{:}5$.
2. $1/125 = 1/5^3 = (12/60)^3 = 1{,}728/60^3$. Use greedy algorithm: $1{,}728 = 28 \times 60 + 48$, therefore $1{,}728/60^3 = 28/60^2 + 48/60^3 = 0.0{:}28{:}48$. Base-10 numbers are used in this calculation because we do not have sexagesimal multiplication tables.

5.2.3 (p. 214): He probably would have divided by the closest regular number, 48 or 50. Perhaps he would have taken the average of these two results as a better approximation.

5.3.1 (p. 221): The number 13 must be either *a* or *b* because it is less than 85. It cannot be *b* because *b* must be an even number. (Why?) Thus $a = 13 = (p^2 - q^2)$ and $c = 85 = (p^2 + q^2)$. Therefore $p^2 = (13 + 85)/2 = 49 = 7^2$, $q^2 = (85 - 13)/2 = 36 = 6^2$, and $b = 2pq = 2 \times 7 \times 6 = 84$. Another way is to use Neugebauer's method: guess that $a = 13$ and $c = 85$: Is there an integer, *b*, such that $b^2 = 85^2 - 13^2 = 7{,}056$? Yes, $b^2 = 84^2$.

5.3.2 (p. 229): The right triangle $(b - a, h, b + a) = (1, h, 7)$ defines *h*, therefore $h = 7^2 - 1 = 6.93$, *seked* $= (b - a)/h = 1/6.93 = 0.144$, slope $= 90° - \tan^{-1}(0.144) = 81.8°$.

5.4.1 (p. 239): Using *pq* theory, one way is simply to insert $p/q = 1 + \sqrt{2}$ into the *triples* expression for a/b and show that it yields $a/b = 1$. A more elegant way is to let $a/b = (p/q - q/p)/2 = 1$ and solve for p/q: Let $r = p/q$, so that $(p/q - q/p)/2 = 1$ can be rewritten as $r(r - 2) = 1$. One way of solving this is by completing the square with the cut-and-paste geometric algebra: the rectangle of area $r(r-2) = 1$ is converted into a *gnomon* whose area is the difference between the areas of two squares, one of area $(r-1)^2$, the other of area 1 so that $1 = (r-1)^2 - 1$ or $(r-1)^2 = 2$. Taking the square root of both sides of the equation yields $r = p/q = 1 + \sqrt{2}$, QED.

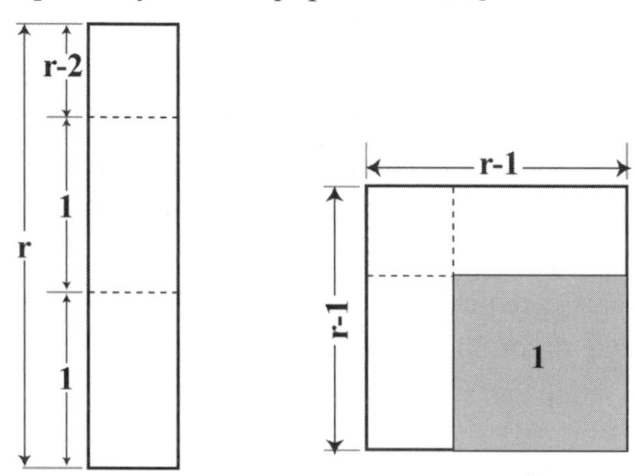

Figure FQ Answer 5.4.1a

Note how this cut-and paste solution is equivalent to the multiplication method of equation (5.1.1a): multiplication is visualized as a rectangle whose area is given as the difference between two squares. The cut-and-paste recipe for converting a rectangle into a gnomon is

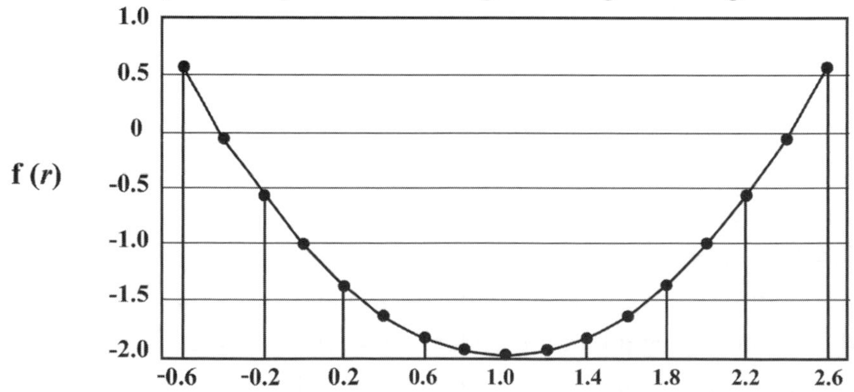

Figure FQ Answer 5.4.1b

a more sophisticated method than starting with a prepartitioned square, as illustrated, for example, by any of the figures in section 5.1, and was probably a later geometric-algebra development.

Diagrammatic visualizations of solutions to quadratic equations are still useful, but the modern method is rather different: Let $f(r) = r^2 - 2r - 1$ [read f(*r*) as "function of *r*]." Graph f(*r*) vs. *r* as illustrated. The solutions are where f(*r*) crosses the *r* axis. There is a solution for $r = 1 + \sqrt{2}$, just as obtained by OB geometric algebra. There is also a solution for $r = 1 - \sqrt{2}$, which the Babylonians could not have obtained, nor would negative measures have had any meaning for them. OB geometric algebra is clearly less general than modern graphical conceptions. But it is of great historical interest as the starting point of quadratic algebra.

5.4.2 (p. 239): $a/b = b/(a + b)$ can be rearranged as: $(a/b)^2 + (a/b) = 1$. Letting $r = a/b$, the quadratic equation can be rewritten more simply as $r(r + 1) - 1$. Using cut-and-paste geometric algebra:

Figure FQ Answer 5.4.2a

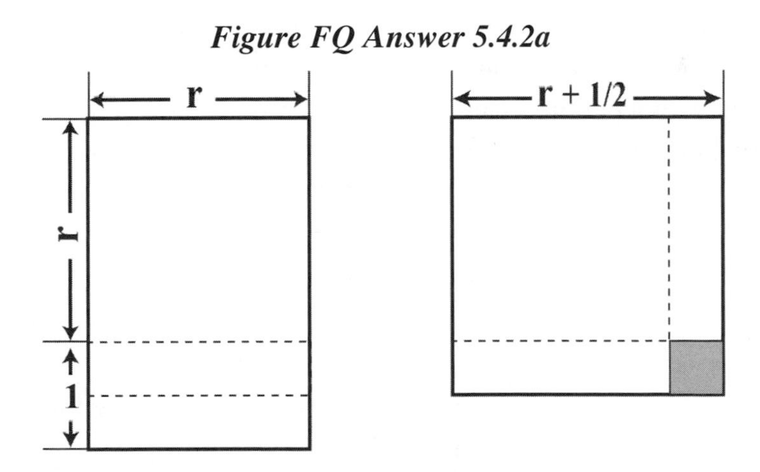

the rectangle of area $r(r + 1) = 1$ is converted into a *gnomon*. The square of area $(r + 1/2)^2$ is completed by adding an area of $(1/2)^2$ (shaded). Thus $(r + 1/2)^2 = 1 + (1/2)^2$, which yields $r = a/b - (\sqrt{5} -$

1)/2, QED. Babylonians or Hindus may have been the first to design buildings according to numerological rules, but they were not the last.

Since $r + 1 = 1/r$ and $r = (\sqrt{5} - 1)/2 = 0.618\ldots$, therefore $r + 1 = 1.618\ldots$. This is a number that crops up frequently in mathematics and is given the Greek letter ϕ (phi) = 1.618. . . . If the Pyramids of Giza had been designed according to a "golden section" criterion as in the drawing,

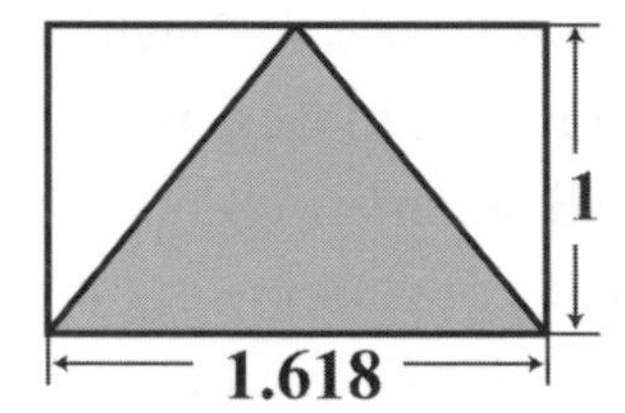

Figure FQ Answer 5.4.2b

the slope of a pyramid would be defined by a *seked* = $\phi/2 = 0.81$. Referring to table 4.4.1, this exactly gives the slope of the Pyramid of Mycerinus. This coincidence has not gone unnoticed by the school of mystical significance of pyramid dimensions.

The Washington Monument in Washington, DC, is an obelisk capped with a pyramid with dimensions also defined approximately by ϕ:

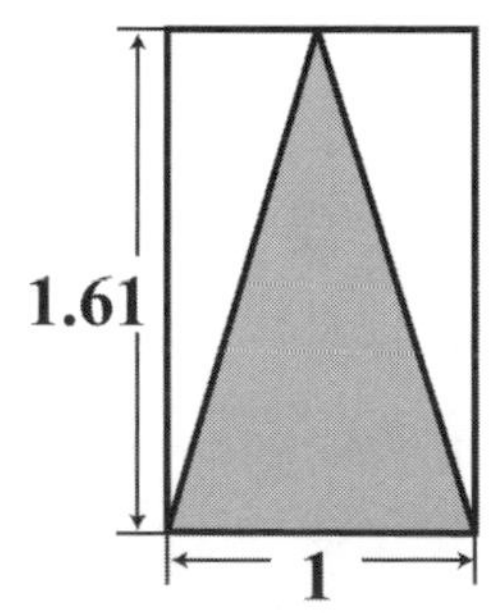

Figure FQ Answer 5.4.2c

Whether this is a planned "golden section" or just a coincidence is not a question that interests me.

5.4.3 (p. 240): YBC 6967 defines a rectangle with sides $a = 5$, $b = 12$. $5^2 + 12^2 = 13^2$ (note that here the numbers are the same whether expressed in base-10 or in transcribed base-60 because numbers from 1 to 59 are written the same in both systems), a Pythagorean triple, which can be accounted for by *pq* theory with $p = 3$ and $q = 2$. It doesn't appear in Plimpton 322 because $p/q = 1.5_{10}$ is not in the p/q range covered by the table.

Db_2 146 defines a rectangle with sides $a = 0.45_{60}$, $b = 1.0_{60}$, and c (the diagonal) $= 1.15_{60}$, which can be rewritten as: $a = 3 \times 0.15_{60}$, $b = 4 \times 0.15_{60}$, and $c = 5 \times 0.15_{60}$, QED.

5.5.1 (p. 245): We know that $c_0 = 2$ is too small because $c_0^2 = 4 < 5$. We know that $c_0 = 3$ is too big because $c_0^2 = 9 > 5$. Thus a reasonable first guess is $c_0 = 2.5$, and hence $c_1 = (c_0 + 5/c_0)/2 = 2.25$; $c_2 = 2.2361111$; $c_3 = 2.236068$. The correct value is 2.236068, to seven-decimal-place precision, and we have obtained exactly this value with only three iterations!

6.1.1 (p. 250): The line-of-sight distance to the horizon is effectively the same as the distance along the slightly curved water surface, so by the Pythagorean theorem: $R^2 + d^2 = (R + h)^2 = R^2 + 2hR + h^2$. Since $R >> h$, the term h^2 is negligible so that $d \cong \sqrt{2hR}$. The diameter of the earth is $2R = 40 \times 10^6/\pi = 12.7 \times 10^6$ m, and $d \cong \sqrt{12.7 \times 10^4}$ m = 5 km. If, however, you were to stand in the water with your eyes just 1 cm above sea level, with $h = 10^{-2}$ m, then $d \cong \sqrt{25.4 \times 10^6}$ = 356 m. You can measure such a distance with a string. Thus, Pythagoras could have measured the diameter of the earth: eyes 1 cm above the water, measure distance to the horizon as 356 m, and calculate the diameter of the earth as 12,700 km. The Pythagorean theorem is a surprisingly useful.

You may feel that this derivation is too sophisticated to have been done by about 500 BCE. The critical step is the appreciation that in the expansion, $(R + h)^2 = R^2 + 2hR + h^2$, the term h^2 can be neglected, but this is just the insight the Babylonians had at least some one thousand years earlier in their square root calculation, $(c_0 + e_0)^2\ c_0{}^2 + 2c_0e_0$ (see section 5.3, figure 5.3.5).

6.1.2 (p. 251): $D + d \cong \sqrt{2R}\ (\sqrt{h} + \sqrt{H}) = 3{,}560(\sqrt{2} + \sqrt{4{,}000}) =$ 230 km.

6.1.3 (p. 254): In the sum of the series of integers: $S = 1 + 2 + \ldots + n - n\ (n + 1)/2$, let $n = 2m$, and obtain the sum up to some even number, $2m$: $S = m(2m + 1)$. From this sum subtract the sum for odd numbers up to the odd number $2m - 1$: $S_o = 1 + 3 + \ldots + (2m - 1) = m^2$ and obtain the sum for even numbers up to $2m$: $S_e = S - S_o = m(2m + 1) - m^2 = m(m + 1)$.

6.2.1 (p. 256): $\sqrt{500} = 22.4$. From table 6.2.1, the primes less than 22 are 2, 3, 5, 7, 11, 13, 17, and 19.

6.2.3 (p. 258): March 22 is the day of the vernal equinox when nighttime and daytime durations are equal. At high noon on this day, the sun is directly overhead at the equator; therefore, your latitude is 45° north of the equator. In case it is raining on March 22, you can repeat the measurement on September 22, the day of the autumnal equinox, and you will get the same answer.

6.3.1 (p. 260): Let m and n be any integers and thus $2m$ and $2n$ are both even numbers and $2m + 1$ and $2n + 1$ are both odd numbers:

$$2m(2n) = 2j, \text{ where } j = 2mn \text{ is an integer, and}$$
$$(2m + 1)(2n + 1) = 4mn + 2m + 2n + 1 = 2k + 1,$$
$$\text{where } k = 2mn + m + n \text{ is an integer, QED.}$$

7.1.1 (p. 268): The following algorithm is not exactly the same as one once learned by many children, but it uses the same concept and its logic is more transparent.

$$(A_2A_1A_0)^2 = (10^2A_2 + 10A_1 + A_0)^2$$

$$(A_2A_1A_0)^2 = [10^4 A_2^2] + [10^3(2A_2A_1) + 10^2(A_1^2)] + [10^2(2A_2A_0) + 10(2A_1A_0) + A_0^2]$$

To calculate $\sqrt{123{,}456}$, let $(A_2A_1A_0)^2 \cong 123{,}456$.Choose the greediest A_2 such that the first bracketed term is less than or equal to 123,456.

$$A_2 = 3;\ 10^4A_2^2 = 90{,}000;\ 123{,}456 - 90{,}000 = 33{,}456$$

Choose the greediest A_1 such that the second bracketed term is less than or equal to 33,456.

$$A_1 = 5;\ [10^3(2A_2A_1) + 10^2(A_1^2)] = 32{,}500;\ 33{,}456 - 32{,}500 = 956$$

Choose the greediest A_0 such that the third bracketed term is less than or equal to 956.

$$A_0 = 1;\ [10^2(2A_2A_0) + 10(2A_1A_0) + A_0^2] = 701;\ 956 - 701 = 255 = \text{remainder}$$

Check answer: $(351)^2 + 255 = 123{,}456$

To calculate $\sqrt{12{,}345.6}$, let $(A_2A_1A_0)^2 \cong 12{,}345.6$

$$A_2 = 1;\ 10^4A_2^2 = 10{,}000;\ 12{,}345.6 - 10{,}000 = 2{,}345.6$$

$$A_1 = 1;\ [10^3(2A_2A_1) + 10^2(A_1^2)] = 2{,}100;\ 2{,}345.6 - 2{,}100 = 245.6$$

$A_0 = 1$; $[10^2(2A_2A_0) + 10(2A_1A_0) + A_0^2] = 221$; $245.6 - 221 = 24.6$
$=$ remainder

Check answer: $(111)^2 + 24.6 = 12{,}345.6$

To calculate $\sqrt{1{,}234.56}$, note that $123{,}456 = 10^2(1{,}2345.56)$, so $\sqrt{123{,}456}$, $= 10 \times \sqrt{1{,}234.56}$ and therefore $\sqrt{1{,}234.56} \cong 35.1$. Generalizing this result, if $\sqrt{N_0}$ is known, then for other numbers $N = 10^{2n}N_0$, $\sqrt{N} = 10^n$, $\sqrt{N_0}$ where n = integer. In those primitive years before the invention of the electronic calculator when square root tables were used, this result allowed a finite, practical number of table entries to cover an infinite range of *N*s.

Learning this algorithm had the marginal value of showing children that square roots were calculable by other than just by judicious guessing, although that is also a valid but somewhat more tedious method.

NOTES AND REFERENCES

References are listed alphabetically by author's last name for each chapter. Internet sites for which no author is given are listed under www. Most references are chosen to be readable, accessible, and have extensive bibliographies to the original scholarly studies so that any data or concept can be traced back to an original source. Some references are to recent scholarly works not yet referenced in readily accessible books or journals.

Nowadays the Internet is by far the most readily available source, and I give the Web sites of some Internet references. One problem with a Web site reference is that when sponsorship ceases, so does the Web site. By the time this book is published, some of the listed Web sites will probably no longer exist. Another problem is that a Web site can be edited anytime by its sponsor without any reference to the changes made so that a fact or idea referenced as from a Web site may not be found in the current version. (The author of a book may also change his/her mind, but the change is then denoted as a new edition.) Thus, although the Internet is a useful information source, a Web site reference is not always verifiable even when the

Web site still exists. Rather than relying on Web site addresses, it is better to use an Internet search engine (Google and the like) with keywords such as "Babylonian Mathematics," or "Pythagoras," or whatever. The search engine will list relevant Web sites according to some measure of popularity, and therefore somewhat in order of usefulness.

1. INTRODUCTION

Boyer, C. B., and U. C. Merzbach. 1991. ***A History of Mathematics*** **(New York: Wiley, first printed in 1968).** More about the life and works of any mathematician I mention can be found here. It also contains a brief review of pre-Greek mathematics.

Gullberg, J. 1996. ***Mathematics from the Birth of Numbers*** **(New York: Norton)** is a somewhat unconventional source, a general mathematics reference with historical commentary. He nicely illustrates the hands-plus-toes origin of base-20 number systems.

Hofstadter, R. 1955. ***Social Darwinism in American Thought*** **(Boston: Beacon, first printed in 1944)** reviews implications of the theory of evolution on fields other than biology, but does not include mathematics. Social Darwinism has a somewhat tarnished reputation, not because the concept is questionable, but because people have misused it to support their political agendas.

Koza, J. R., et al. 2003. "Computer Programs That Function via Darwinian Evolution Are Creating Inventions That Are Novel and Useful Enough to Be Patented." ***Scientific American*** **288, no. 6: 40** is a popularized account, in a magazine available in most libraries, by the developers of genetic programming.

Newman, J. R. 1956. ***The World of Mathematics*, Volume 1, Part III:** ***Arithmetic, Numbers, and the Art of Counting*** **(New York: Simon & Schuster)** is a collection of articles written by noted mathematicians with many articles relevant to the childhood of mathematics.

The selection by L. L. Conant, "Counting," gives examples of pebble counting with 1-for-10 replacements.

O'Connor, J. J., and E. F. Robertson maintain the Web site of the MacTutor History of Mathematics Archive, www.history.mcs.st-andrews.ac.uk/history/index. This is a handy source for biographies of mathematicians and the history of ancient mathematics.

Sarton, G. 1993. *Ancient Science through the Golden Age in Greece* (New York: Dover, first printed in 1952) starts with an extensive review of pre-Greek mathematics.

2. The Birth of Arithmetic

2.1 Pattern Recognition Evolves into Counting

Leakey, M., and A. Walker. 1997. "Early Hominid Fossils from Africa." *Scientific American* 276, no. 6: 60–65 is a popularized account, in a magazine available in most libraries, by Meave Leakey, the discoverer of important ancient hominid finds in Africa.

Newman, J. R. 1956. *The World of Mathematics*, Volume 1, Part III: *Arithmetic, Numbers, and the Art of Counting* (New York: Simon & Schuster). The selection by O. Koehler, "The Ability of Birds to 'Count,'" is particularly relevant to pattern recognition vs. counting.

Wong, K. 2003. "An Ancestor to Call Our Own." *Scientific American* 288, no. 1: 42 reports, in a magazine available in most libraries, on recent discoveries by Michel Brunet of seven-million-year-old hominid remains in Chad.

2.2 Counting in Hunter-Gatherer Cultures

Brooks, A. S., and C. Smith. 1987. "Ishango Revisited: New Age Determinations and Cultural Interpretations." *African Archaeo-*

logical Review **5: 67–78** dates the Ishango Bone as about twenty thousand years old.

Closs, M. P. 1996. ***Native American Mathematics*** **(Austin: University of Texas Press)** is my source for recent analyses of Native American mathematics.

Heinzelin, J. de. 1962. "Ishango." ***Scientific American*** **206, no. 6: 105–16** is an article, in a magazine available in most libraries, by the discoverer of the Ishango Bone. While this article dates the bone as eighty-five hundred years old, the more recent dating by Brooks and Smith (cited above) is presumably more reliable and that is the result I have quoted. However, a very similar notched bone dated at thirty thousand years ago was found in Czechoslovakia in 1937 by anthropologist Karl Absolom. Wall marks in a cave in Lascaux, France, dating from twenty thousand years ago clearly show counting days in a lunar cycle. Thus, whatever is the the true dating of the Ishango Bone, it is rather certain that humans were able to count to higher than ten at least by about twenty or thitry thousand years ago even though they may not have had words to express large numbers.

Ifrah, G. 1998. ***The Universal History of Numbers*** **(New York: Wiley, first printed in 1994)** is a scholarly work of encyclopedic coverage of the beginnings of counting, yet it is very readable.

Joseph, G. G. 2000. ***The Crest of the Peacock: The Non-European Roots of Mathematics*** **(Princeton, NJ: Princeton University Press, first printed in 1991)** is also a source for information on the Ishango Bone. The uncritical acceptance of dubious interpretations of the Ishango Bone is a flaw in an otherwise useful and readable book.

Pinker, S. 1999. ***Words and Rules: The Ingredients of Language*** **(New York: Basic Books)** is the basis of my *words-and-rules* analysis of hunter-gatherer counting.

Schmandt-Besserat, D. 1996. ***How Writing Came About*** **(Austin: University of Texas Press)** is an abridged version of her ***Before Writing*** (first printed in 1992). My references to concrete counting are from here.

Williams, S. W. maintains the ethnomathematics Web site, Mathematicians of the African Diaspora, www.math.buffalo.edu/mad/ancient-africa. He discusses the Ishango Bone.

3. Pebble Counting Evolves into Written Numbers

3.1 Herder-Farmer and Urban Cultures in the Valley of the Nile

Gillings, R. J. 1982. *Mathematics in the Time of the Pharaohs* (New York: Dover, first printed in 1972) is a popularized, but very detailed description of ancient Egyptian mathematics.

Joseph, G. G. 2000. *The Crest of the Peacock: The Non-European Roots of Mathematics* (Princeton, NJ: Princeton University Press, first printed in 1991) has an excellent chapter, "The Beginnings of Written Mathematics: Egypt."

O'Connor, J. J., and E. F. Robertson review Egyptian mathematics on the Web site of the MacTutor History of Mathematics Archive, www.history.mcs.st-andrews.ac.uk/history/index.

Singh, S. 1999. *The Code Book* (New York: Doubleday). Singh gives a brief history of the deciphering of Egyptian hieroglyphics. Particularly interesting about Singh's treatment is that it is given in the more general context of deciphering any code, including military and Internet encryption.

www.petrie.ucl.ac.uk/digital_egypt/weights/area is part of the Web site of the Petrie Museum of Egyptian Archeology, University College London. From the confusing assortment of Egyptian measurement units, which have changed names and values during their centuries of use, this site has been able to identify their lineage.

3.2 Herder-Farmer and Urban Cultures by the Waters of Babylon

Friberg, J. 1985. "Numbers and Measures in the Earliest Written Records." *Scientific American* 250, no. 2: 78–85 is a popularized account, in a magazine available in most libraries, by a scholar involved in deciphering and interpreting Babylonian clay counters (tokens).

Gazale, M. 2000. *NUMBER: From Ahmes to Cantor* (Princeton, NJ: Princeton University Press) is one of few references that recognize

Babylonian numbers as more properly described as an alternating sequence of 1-for-10 and 1-for-6 replacements rather than simply as base-60.

Ifrah, G. 1998. ***The Universal History of Numbers*** **(New York: Wiley, first printed in 1994)** is the source for most of the Babylonian, pebble-counting data I have cited. Although we are both looking at the same data, our interpretations of the origin of base-60 are completely different.

Joseph, G. G. 2000. ***The Crest of the Peacock: The Non-European Roots of Mathematics*** **(Princeton, NJ: Princeton University Press, first printed in 1991)** uncritically repeats the conventional and wrong interpretation of the Babylonian numbers as simply base-60.

O'Connor, J. J., and E. F. Robertson discuss Babylonian base-60 on the Web site of the MacTutor History of Mathematics Archive, www.history.mcs.st-andrews.ac.uk/history/index.

Ore, O. 1988. ***Number Theory and Its History*** **(New York: Dover, first printed in 1948)** hypothesizes that the origin of Babylonian base-60 was weighing units. Although I have rejected this explanation, Ore was possibly the first to perceive that the origin of base-60 was to make the number system conform to measurement units.

Schmandt-Basserat, D. 1978. "The Earliest Precursor of Writing." ***Scientific American*** **238, no. 6: 50–58** is a much abbreviated, popularized account in a magazine available in most libraries.

———. 1996. ***How Writing Came About*** **(Austin: University of Texas Press)** is the source for the early pebble-counting (she refers to them as tokens) data in the Middle East. A token in the shape of a six-pointed star is cataloged and sketched in this book as entry 3.75 in her chapter "The Artifacts." There is no way of knowing whether this was just another distinguishable shape or whether there was some significance to its display of the number six. Although I do not accept her conclusion about concrete counting, this is a very readable account by the discovering archeologist herself.

www.jewishencyclopedia.com describes biblical weights and measures, which correspond with Babylonian ones. To find this information use

SEARCH: "weights" on this Web site. The ancient, dual use of the same units for seed volume and land area is also referenced here.

www.menorah.org/starofdavid.html is the Web site address of my source for the history of the Star of David.

www.wikipedia.org/wiki/historical_weights_and_measures gives a summary of measurement units for many cultures.

3.3 In the Jungles of the Maya

Campbell, J. 1972. *The Hero with a Thousand Faces* (Princeton, NJ: Princeton University Press, first printed in 1949) exhaustively documents the theory that all religions essentially tell the same story. His *monomyth* theory states that religious myths are like dreams; they are subconscious expressions of universal fears and desires. Linguists (see section 2.2, Pinker reference) have discovered that all languages are structured essentially alike, but they have gone a step beyond Campbell and have begun considering language universality in terms of brain circuitry. It is probably fair to say that the way we count and do arithmetic is also because of the way our brains are wired. Sometimes we get a glimpse at how differently wired brains do arithmetic. Autistic people, although lacking ability to function socially, are sometimes capable of remarkable arithmetic. The 1988 movie *Rain Man* depicted (realistically by Hollywood standards) the behavior of such an autistic *savant*.

Culbert, T. P. 1993. *Maya Civilization* (Washington, DC: Smithsonian Books) is a brief but up-to-date review of Mayan archeological discoveries.

Finkelstein, I., and N. A. Silberman. 2001. *The Bible Unearthed: Archeology's New Vision of Ancient Israel and the Origin of Its Sacred Texts* (New York: Free Press). I have adopted this recent interpretation and chronology of biblical archeology. In the last thirty or so years, understanding of the biblical period in Canaan has remarkably improved. However, it is doubtful that there will ever be complete agreement on a subject that is so emotionally charged for so many people. My example of migrations in the Middle East under Ottoman rule is also from this source.

Gazale, M. 2000. ***NUMBER: From Ahmes to Cantor*** **(Princeton, NJ: Princeton University Press)** is a recent reference to a book by a competent mathematician to show that my "obvious" conclusion that Mayan numbering for astronomy is simply measuring with mixed units has not been generally perceived. Gazale's description of Mayan numbering as having "a strange singularity in their number system" has been the accepted interpretation for about a century.

Joseph, G. G. 2000. ***The Crest of the Peacock: The Non-European Roots of Mathematics*** **(Princeton, NJ: Princeton University Press, first printed in 1991)** uncritically repeats the classic interpretation of the Mayan number system.

O'Connor, J. J., and E. F. Robertson review Mayan mathematics on the Web site of the MacTutor History of Mathematics Archive, www.history.mcs.st-andrews.ac.uk/history/index.

4. MATHEMATICS IN THE VALLEY OF THE NILE

Allen, D. maintains a Web site for the notes of a course on the history of mathematics given at Texas A&M University, www.math.tamu.edu/~don.allen/masters/hist_frame.htm. Many college-level lecturers put their class notes on the Internet.

Chance, A. B., et al. 1927. ***The Rhind Mathematical Papyrus.*** **2 Volumes (Oberlin, OH: Mathematical Association of America)** is the classical translation on which are based all modern studies of ancient Egyptian mathematics.

Däniken, E. von. 1996. ***The Eyes of the Sphinx*** **(New York: Berkley).** In the words of the author: "The newest evidence of extraterrestrial contact in ancient Egypt."

Eppstein, D. www.ics.uci.edu/~eppstein/numth/egypt/. This is a Web site on Egyptian fractions with links to sophisticated material for number-theory mathematicians, and to material understandable and of interest to a general reader. Prof. Eppstein kindly traced for me the coinage of the expression *greedy algorithm* back to publications by mathematicians in 1971, but he noted that there probably was earlier use. *Greedy algorithm* sounds like computer-era jargon, and so my guess is that the expression was not coined until about mid-twentieth century.

Friberg, J. 2005. *Unexpected Links between Egyptian and Babylonian Mathematics* (Singapore: World Scientific Publishing) is a recent book that I have not yet read. However, the author has kindly reviewed my notes on the subject and has informed me that we are not relying on the same evidence to arrive at the same conclusion. I am indebted to correspondence with Prof. Friberg for improving my understanding of Babylonian mathematics and absolve him of any responsibility for any residual misunderstanding on my part.

Gillings, R. J. 1982. *Mathematics in the Time of the Pharaohs* (New York: Dover, first printed in 1972) covers every extant ancient Egyptian mathematical text. Although he gives the literature reference to Sylvester's conjecture that Egyptians calculated unit fractions by Fibonacci's greedy-algorithm method, he apparently did not understand it because he makes no use of it or even comments about it. Gillings presents derivations of the volume of a frustum by geometric reasoning. He is also the source of the pyramid data in my table 4.4.1. In his appendix 5, he collects the opinions of various historians who unanimously state that there is no evidence that the Egyptians were aware of either the Pythagorean theorem or Pythagorean triples. This appendix creates an unwarranted perception of certainty because all of the historians were reading from the same few surviving documents. The sparseness of the Egyptian mathematical record allows no such certainty.

Joseph, G. G. 2000. *The Crest of the Peacock: The Non-European Roots of Mathematics* (Princeton, NJ: Princeton University Press, first printed in 1991) has a very readable chapter, "The Beginnings of Written Mathematics: Egypt." His chapter, "Ancient Indian Mathematics," is my source for information on the Harappan culture.

Knott, R. www.mcs.surrey.ac.uk/personal/r.knott/fractions/egyptian/.html. This is a Web site on Egyptian fractions. I know of no print publications on the history of Egyptian mathematics that relate to the greedy algorithm although its importance in understanding Egyptian fractions is appreciated by Knott and other mathematicians.

Legon, J. www.legon.co.uk/gizaplan. This is the Web site of John Legon, amateur Egyptologist and discoverer of the role of $\sqrt{2}$ and $\sqrt{3}$ in the Giza plan. I am indebted to him for correspondence that helped me understand his concept.

O'Connor, J. J., and E. F. Robertson review Egyptian mathematics on the Web site of the MacTutor History of Mathematics Archive, www.history.mcs.st-andrews.ac.uk/history/index.

Waterfield, R. 1998. ***The Histories/Herodotus*** **(Oxford: Oxford University Press)**. My quote from Herodotus (Book II, 109) is from this very readable, new translation into modern English.

www.greatpyramid.net is the address of a Web site that claims that the Great Pyramid is "the revelation of Adam in stone." Piazza Smyth is still alive and well on the Internet in the twenty-first century!

5. MATHEMATICS BY THE WATERS OF BABYLON

Allen, D. Has a section on Babylonian mathematics on his Web site, www.math.tamu.edu/~don.allen/masters/hist_frame.htm.

Fowler, D., and E. Robson. 1998. "Babylonian Square Roots: YBC 7289 in Context." ***Historica Mathematica*** **25: 366–78** presents a cut-and-paste calculation of $\sqrt{2}$. I present a slightly modified version of their treatment.

Hoyrup, J. 1996. ***History of Science*** **34: 1** is a shorter and easier-to-read summary of Hoyrup's cut-and-paste interpretation of Babylonian problem texts.

———. 2002. ***LENGTHS—WIDTHS—SURFACES*** **(New York: Springer).** This scholarly book of encyclopedic scope updates the classic work of Neugebauer. Almost all of the cuneiform tablets that I have discussed are referenced in Hoyrup's book.

He is the originator of the cut-and-paste interpretation of some Babylonian problem texts and is the principal advocate of a geometric-algebra interpretation rather than Neugebauer's purely algebraic interpretation. He credits Evert Bruins as the first to realize (1982) that Babylonian algebra is really geometric algebra.

Hoyrup devotes a complete chapter, "The Origin of Old Babylonian Algebra," to why the problem texts were composed. He ignores the O'Connor and Robertson conjecture that Babylonian multiplication related to the same algorithms that appear in the problem texts. I am indebted to correspondence with Professor Hoyrup for improving my understanding of Babylonian mathematics and absolve him of any responsibility for any residual misunderstanding on my part.

———. 2002. "A Note on Old Babylonian Computational Techniques." ***Historica Mathematica*** **29: 193–98** is an attempt to understand an OB multiplication by analysis of errors in the calculation. He shows a possible solution that is consistent with my conjecture about the use of multiplication-table memorization.

Joseph, G. G. 2000. ***The Crest of the Peacock: The Non-European Roots of Mathematics*** **(Princeton, NJ: Princeton University Press, first printed in 1991)** has a very readable chapter, "The Beginnings of Written Mathematics: Babylon." Joseph uncritically accepts the purely algebraic derivation of Babylonian algorithms. His chapter "Ancient Indian Mathematics" is my source for information on Hindu mathematics.

Melville, D. J. http://it.stlawu.edu/~melvill/mesomath/ is the Web site for notes of a History of Mathematics course given at St. Lawrence University. I am also indebted to Professor Melville's helpful correspondence on VAT 6505, OB reciprocal tables, and OB multiplication. I absolve him of any responsibility for any residual misunderstanding on my part.

Neugebauer, O. 1962. ***The Exact Sciences in Antiquity*** **(New York:**

Dover, first published in 1952) is a more readily available summary of some of Neugebauer's studies.

Neugebauer, O., and A. Sachs. 1945. ***Mathematical Cuneiform Texts*** **(New Haven: American Oriental Society and the American Schools of Oriental Research),** a work of remarkable scholarship, is the classic translation and analysis of Babylonian, mathematical clay tablets on which all subsequent studies are based.

O'Connor, J. J., and E. F. Robertson review Babylonian mathematics on the Web site of the MacTutor History of Mathematics Archive, www.history.mcs.st-andrews.ac.uk/history/index. This is the source of the Babylonian tables-of-squares theory of Babylonian multiplication. O'Connor and Robertson categorically assert that the Babylonians used multiplication algorithms based on tables of squares. Since I know of no direct epigraphic evidence, I presume that the assertion is actually an insightful conjecture based on the indirect evidence that problem texts relate to the same equations, as do their multiplication algorithms. These authors do not relate Babylonian multiplication algorithms to geometric visualization nor do they imply that the multiplication algorithms are the antecedents of Babylonian algebra; these are my conjectures.

Ore, O. 1988. ***Number Theory and Its History*** **(New York: Dover, first printed in 1948)** explains Plimpton 322 in terms of *regular numbers*.

Robson, E. 1997. "Three Old Babylonian Methods for Dealing with 'Pythagorean' Triangles." ***Journal of Cuneiform Studies*** **49: 51** shows that although OB scribes had apparently discovered Newton's method for finding a square root, they did not completely understand it and usually employed much poorer approximations for other than the square root of two.

———. 2001. "Neither Sherlock Holmes nor Babylon: A Reassessment of Plimpton 322." ***Historia Mathematica*** **28: 167–206** is a reference that sums up many of the previous studies of Plimpton 322. I am indebted to correspondence with Dr. Robson for improving my understanding of Babylonian mathematics and absolve her of any responsibility for any residual misunderstanding on my part.

Sachs, A. J. 1947. "Babylonian Mathematical Texts I: Reciprocals of Regular Sexagesimal Numbers." ***Journal of Cuneiform Studies*** **1: 219–40** is his translation of VAT 6505.

Singh, S. 1998. ***Fermat's Enigma*** **(New York: Walker)** is the story of the trail from Pythagoras to Diophantus to Pierre de Fermat to Andrew Wiles. It now appears that the trail started some two thousand years earlier in Babylon with Plimpton 322.

Strom, A. (1877–1951). The education of Abe Strom, my grandfather, was in a Yeshiva in a village in Lithuania. His formal mathematics education was minimal, yet he was able to rapidly solve word problems equivalent to the quadratic algebra of YBC 6967 ($xy = 60$, $y - x = 7$). He realized that the solution must be a pair of factors of 60 and he mentally sequenced through the candidates until he came to the pair, (5, 12), where the difference between the factors was 7. He never learned about factoring, prime numbers, and other number concepts, but understood many of them intuitively. He was "number smart," as were many in his pre–electronic calculator era.

I reference Abe Strom's method because Babylonian scribes could have used the same method four thousand years ago. It is by far the fastest way to solve this problem. That they did not use it is evidence that their problem texts were not recreational math as we saw in section 4.3 was Egyptian practice, but were exercises in geometric algebra.

Waerden, B. L. van der. 1983. ***Geometry and Algebra in Ancient Civilizations*** **(New York: Springer-Verlag)** theorizes that Pythagorean triples were important for ancient religious rituals from Stonehenge to China. I find that difficult to believe, but after learning of ancient Hindu use, I am no longer so skeptical.

www.cut-the-knot.com is the Web site of Cut-The-Knot Inc., a software development company. The site posts comments and questions about Pythagorean triples. An Internet keyword search for "Pythagorean theorem" will turn up tens of Web sites with Pythagorean theorem proofs. Many of these present original, *animated-geometry* proofs that are both amusing and instructive. Would Euclid consider such proofs valid?

6. MATHEMATICS ATTAINS MATURITY: RIGOROUS PROOF

Heath, T. L. 1956. ***Thirteen Books of Euclid's The Elements*** **(New York: Dover, first printed in 1908)** is the modern, standard translation of Euclid and includes very interesting historic commentary. Noteworthy is that there is no mention whatsoever of previous Babylonian mathematics. It was completely unappreciated at the time of the writing.

Mankiewicz, R. 2000. ***The Story of Mathematics*** **(Princeton, NJ: Princeton University Press)** is a review of mathematics from the Ishango Bone to fractals—and with hardly an equation. Nothing is covered in much detail, but it nicely puts mathematical developments in historical sequence.

Sarton, G. 1993. ***Ancient Science through the Golden Age in Greece*** **(New York: Dover, first printed in 1952)** is an encyclopedic but very readable source for Greek mathematics. It nicely puts the results in perspective for all Greek intellectual history.

7. WE LEARN HISTORY TO BE ABLE TO REPEAT IT

7.1 Teaching Mathematics in Ancient Greece and How We Should but Do Not

Levitt, S. D., and S. J. Dubner. 2005. ***Freakonomics*** **(New York: William Morrow)** presents studies that show how and the extent to which teachers in the United States cheat to inflate the grades of their students in standardized tests.

INDEX